the poetics of **dna**

CARY WOLFE, SERIES EDITOR

2 *The Poetics of DNA*
Judith Roof

1 *The Parasite*
Michel Serres

the poetics of dna

judith roof

posthumanities 2

University of Minnesota Press
Minneapolis
London

Published by the University of Minnesota Press
111 Third Avenue South, Suite 290
Minneapolis, MN 55401-2520
http://www.upress.umn.edu

Library of Congress Cataloging-in-Publication Data

Roof, Judith, 1951–
The poetics of DNA / Judith Roof.
p. ; cm. — (Posthumanities ; 2)
Includes bibliographical references and index.
ISBN 978-0-8166-4997-6 (hc : alk. paper) —
ISBN 978-0-8166-4998-3 (pbk. : alk. paper)
1. DNA. 2. Genetics—Social aspects. 3. Developmental genetics.
I. Title. II. Series.
[DNLM: 1. DNA. 2. Genes. 3. Genetic determinism.
4. Symbolism. QU 58.5 R776p 2007]
QP624.R66 2007
572.8′6—dc22

2007007308

Printed in the United States of America on acid-free paper

15 14 13 12 11 10 09 08 07 10 9 8 7 6 5 4 3 2 1

For Fedwa and in memory of Teresa,
the best public intellectuals I have known

Contents

Acknowledgments

This project began as an offshoot of a previous book, *Reproductions of Reproduction.* Its inscription has been spurred by discussions with Fedwa Malti-Douglas and Allen Douglas, Cary Wolfe, Jackie Stacey, Jaime Hovey, Teresa Brennan, Maureen McNeil, Steve Rachman, Patrick O'Donnell, Peter Ford, Brian Olszewski, and Greg Nicholson. I also thank Greg for helping prepare this manuscript.

The book owes its development to such generous venues of discussion as the University of Wyoming, Vanderbilt University, and the University of Lancaster, where I was graciously invited to present this work.

Finally, I thank Richard Morrison, whose enthusiasm brought this to the page.

CHAPTER 1

The Epic Acid

At the gala opening ceremonies of the 2004 Summer Olympic Games in Athens, Greece, organizers mounted a spectacular tableau of Western cultural history. Merging figures of Greek mythology, the changing aesthetics of Hellenic pottery, and an ascending chain of modern humanity, the Olympic pageant metamorphosed from epoch to epoch, culminating in a cosmic Milky Way lake of lights. In a last transformation (effected by a complex system of wires, projections, and laser lights), the luminous lake rose into the air to form the double helix, a spiraling light show representing, if not the alpha, then the omega of it all. The BBC1 commentators who bantered the pageant play-by-play evinced no surprise at the transition from the horizontal cosmos to the whirling vertical of the molecule. It was as if this finale was already anticipated, evincing the "of course" of recognition and familiarity, as if an old friend had arisen from its familiar stellar habitat.

That the opening spectacle of the Olympics would lead to deoxyribonucleic acid seems almost hackneyed in 2004, the transition from cosmos to DNA so natural as to satisfy even epic narrative yearnings (or yearnings for epic narrative occasioned by the Olympics themselves as an epochal sporting event resounding its own history). But how, in the global public fantasy of 2004, do the cosmos and DNA morph from one to the other? What has DNA become that we see it as a cosmic truth, representative of all life, residence of all answers, potential for all cures, repository for all identity, end to all stories?

The DNA figure that made its twinkling appearance at the Olympics is a synecdoche, a part that stands for the whole of life. Although the famous double helix is only a component of a gene, and a gene is only a part of a chromosome, and a chromosome is only one of many in the sum total of a human genotype, and a genotype is only partly responsible for how individuals turn out, the graphic ladder qua "staircase to heaven" has come to stand for it all. This book looks at how discourses about DNA have helped that ascension happen: how the various analogies, metaphors, and other figurations of DNA have introduced a complex set of representational operations that have taken nucleic acid from its sets of complementary components to the answer to all questions. More important, this hyperbolized notion of DNA, as it has become inevitably confused or conflated with our notion of the "gene," has become the vector through which older ways of thinking can merge with the new, through which newer, more threatening ideas can emerge masked by the old, and through which older, more conservative ideas can survive. DNA transmits more than genetic information or life codes. It is more than an evolutionary record of the development of life on earth. In the twenty-first century it has become the symbolic repository of epistemological, ideological, and conceptual change.

Synecdoche, Metaphor, Narrative, Pseudoscience

The completion of the human genome in 2000 accorded with the era's pervasive millennial hype. The "book of life" furnished a fitting climax to four hundred years of Enlightenment science as well as a propitious starting point for the nascent millennium's future labors. The coincidence of millennium and announcement capped the stream of serendipities that had accompanied the last half-century's investigations of DNA-based genetic research. Not only did DNA fit into ideas of "germ" matter promulgated from the Greeks through Charles Darwin (1809–82), it also located the key to life's mysteries in the locus of the very small and structural, the direction biological sciences had been taking since the Enlightenment. James Watson and Francis Crick discerned the self-reproductive structure of DNA in 1953 at the same time that the linguist J. L. Austin was promulgating his theory of the performative, the class of statements that in saying also accomplish what they say.[1] The evolution of DNA-based genetic science accompanied the development of digital computers and theories of cybernetics, and the emergence of the contemporary category of the postmodern. The correlations of thought and molecule suggest that the history of DNA has been a saga of things falling into place, of the happy alignment of ideas and acids. If DNA seems to initiate a cascade of parallels, resonances, and possibilities in the latter half of the twentieth century, it is because DNA was already the climax of a story long in preparation and telling, the anticipated emblem of the era that defined it and that it defined.

Just as this happy climax seems to bring everything together in its enthusiastic comparisons to secret, book, and key, DNA also figures less obvious reactions, compensations, and fictions that preserve structure in the face of more systemic complexity. By subtly importing narrative logics and investments, DNA's analogies encourage a hyperbolic sense of agency and control as well as a host of Western ideologies about identity, gender, and difference. DNA's figurations alibi sets of pseudoscientific ideas that bypass scientific reasoning in favor of magic, instantaneity, and the commodity short-circuiting of work and desire. This pseudoscience may be the inevitable effect of any attempt to represent complex phenomena for a popular audience, but it may also be inherent to the overdetermined sets of analogies, figurations, and unacknowledged narratives that even scientists import to render and occasionally conceptualize their thinking about DNA and genes.

What follows is in part a broad cultural history of the ideas that led to DNA genes, their capabilities, and the mechanisms of their operation. It is also an analysis of how DNA's embodiment of these venerable notions both signals and hides a larger set of epistemological shifts and battles between structure and system, metaphor (substitution) and metonymy (contiguity), hierarchy and plurality, and taxonomic order and entwined complexities. These play out not only through the assumptions associated with various metaphors and figures (language, text, software, homunculus) but also through the even less evident processes of reductionism, narrative, and magical thinking that lead ultimately to pseudoscience. In various ways, those who have written about the representation of genes and the genome have shown that figurations are misleading. This book focuses on the deep structure of these figurations and implicit narratives, showing what kinds of values they import, why those comparisons are valuable, and what they tell us about how we are thinking. Because genetic research and technology continue to proliferate, this book stops at a certain point in time, around 2006. Otherwise, it would never stop at all.

The Synecdochal Gene

DNA, the extolled molecule, is not a gene, but it has come to stand both for the mechanism by which biological information is reproduced and inherited and, at least on the popular front, as the agent of all biological (and sometimes social and cultural) causality. The concept of the gene as an agency that transmits characteristics came well before the term *gene* was coined and eons before the relatively recent description of DNA. In considering the mysteries of reproduction, thinkers from Aristotle to Darwin proposed various mechanisms for how traits were handed from father (or, later, parents) to children. Most of the theories proposed that some form of matter, some particle or "factor," is transmitted from parent to child during reproduction. Darwin called these elements "particles"; Gregor Mendel, the first to identify the rules by which certain traits were transferred, called them "factors." The nature of these factors eluded researchers until the middle of the twentieth century when the biochemist Oswald Avery discerned large amounts of a substance called deoxyribonucleic acid in cell nuclei. DNA became the focus of the researches of the eminent Linus Pauling in California as well as of laboratories in England. Using clues offered by X-ray crystallographic images of DNA molecules made by the London researcher Rosalind Franklin, the

Cambridge graduate fellows Watson and Crick pieced together the necessary configuration of the double helix, with paired nucleic acids forming the inside, or "rungs" of the ladder, and binding sugars forming the sides. Realizing that this structure permitted a continuous set of exact duplications when the nucleic acid rungs parted to create two halves, Watson and Crick hypothesized that DNA's very structure resolved the question of how genetic material made its way from generation to generation.[2]

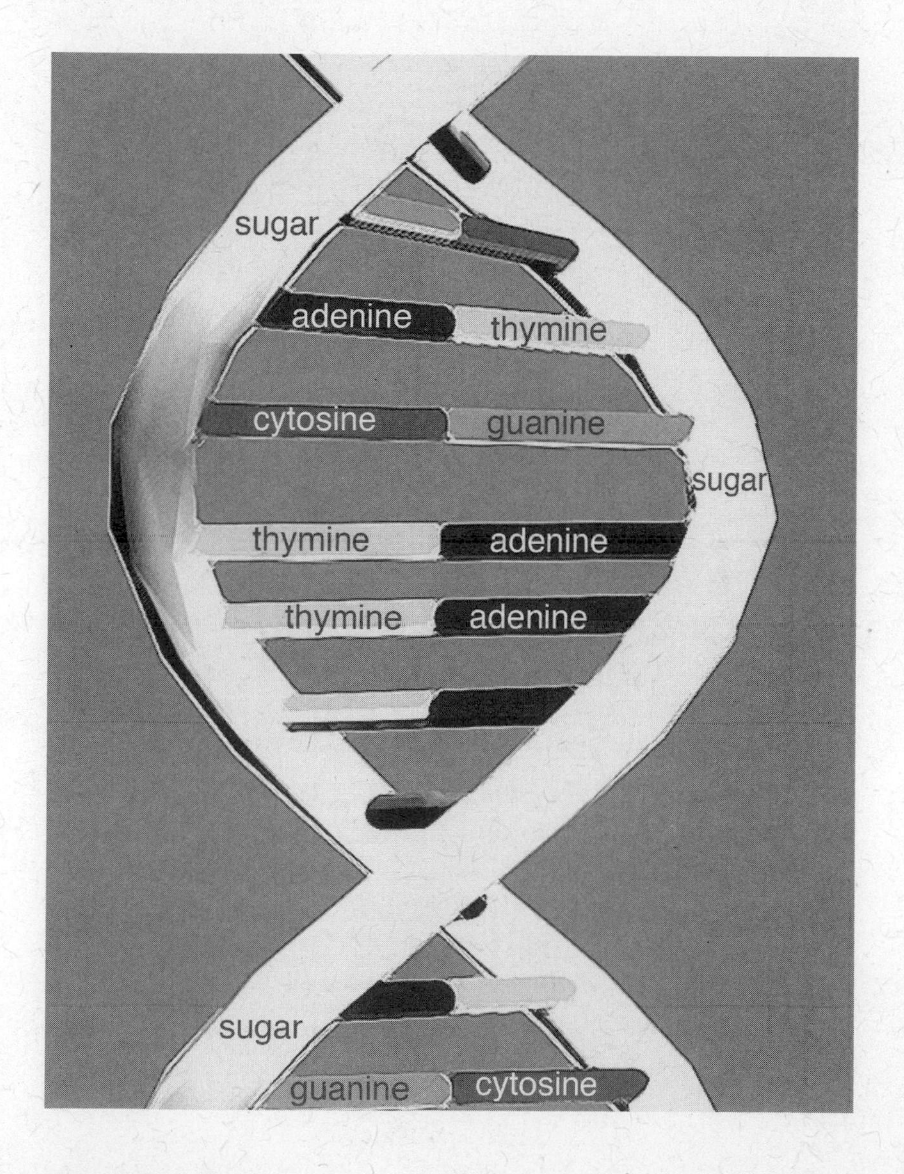

But what exactly is deoxyribonucleic acid? DNA is composed of four different nucleic acids: adenine (A), thymine (T), cytosine (C), and guanine (G). These bases are arranged in pairs like rungs on a ladder, which turns a bit at every rung, producing DNA's famous double helix structure. If the rungs are split apart, each half provides an empty slot that can be occupied only by its complementary opposite. For example, if one side of a DNA ladder consists of A, C, G, T, it can be reconstructed only by inserting the combination T, G, C, A in that order. When the pairs "unzip" and replicate, two identical strands of nucleic acids are produced. In vitro, DNA replicates itself with the help of ribonucleic acid (RNA), which forms a "negative" impression by matching the bases with their complementary opposites (except for thymine, which is replaced on RNA with another base, uracil).

DNA and genes are not the same things. Even though genes are made up of DNA, DNA is not the only element of their functioning. Genes exist in relation to complex environments both large and small, working as part of a complicated set of systems and possibilities. DNA also exists in sites other than cell nuclei, such as in mitochondria, and performs functions other than those associated with heredity, such as governing how life processes proceed within individual organisms. The DNA carried by chromosomes is organized into genes or, as Richard Dawkins defines them, "genetic units," which are "small enough to last for a large number of generations and to be distributed around in the form of many copies."[3] If they afford characteristics for survival, genes may last for eons. Characteristics for survival may include morphological traits such as camouflaging feathers or long legs as well as developmental and chemical processes by which individual organisms survive.

In broader cultural renditions, DNA stands for the gene as its synecdoche, taking over ideas about heredity as well as becoming an immense, vague compendium of "information" believed to cause everything from alcoholism to thrill seeking. This DNA-gene complex is, however, imagined differently from Mendel's idea of the "factor." Mendel's gene factor referred to the transmission of traits, while most contemporary representations of the gene focus on DNA as itself automatically causal. The two categories—the gene as an organized operation,

DNA as a chemical material—have merged conceptually, producing something like a "DNA gene." Redundant as this terminology may seem, it actually represents the overlap of ideas, each part—DNA, gene—contributing a different aspect of a complex of ideas comprising both together.

As the current "starring" half of this DNA-gene composite, DNA is not simply a chemical active in biochemical processes. It stands at the tip of an iceberg of beliefs, ideas, and concepts about how life and science work, what we can do with what we know, and the forms knowledge can take. From the discernment of its structure, DNA has always been more than itself. The concepts DNA has come to represent have appeared at different points in history, mounting and accruing toward mechanisms for heredity, identity, development, cell regulation, and kinship. The three acronymic letters, then, like the chemical itself, have come to signify a vast number of processes, undifferentiated to the nonscientist and rendered intelligible by a series of metaphors or comparisons. These include such analogies as the "secret of life," the code, the book, the alphabet, sentences, words, chapters, histories, the Rosetta stone, the Holy Grail, the recipe, the blueprint, the text, the map, the homunculus, software, and others. None of these analogies is accurate in terms of how DNA works or even what it accomplishes. All of them import values, meanings, mechanisms, and possibilities that are not at all a part of DNA. The effect is that DNA has always stood for much more than what it is.

An early critic of the claims made for genes, Richard Lewontin comments that "there is, at present, no aspect of social or individual life that is not claimed for the genes."[4] Lewontin's corrective commentary, made consistently since the early 1980s, has focused not only on hyperbolic claims made about DNA and genes but also on the hubris of science itself. Working as a scientific counterconsciousness, Lewontin has tried to check the scientific media's overevaluation of scientific progress. Suggesting, for example, that "scientists do what they already know how to do," Lewontin criticizes scientists for sidestepping problems they cannot solve, such as how a "single cell turn[s] into a mouse," in favor of those "they do know how to answer."[5]

In a curious fashion Lewontin's critiques of contemporary science culture also endow science with a great deal of power. His complaint is not about science itself but about the public promulgations of a wrong version of science—a quick, easy, glib science full of hype and fallacious shortcuts in representing what it does do and the actual nature of its discoveries. Lewontin, though somewhat indiscriminate in the eclectic genre of his targets—biology, sociology, psychology—aims at the misperceptions at the core of research into questions scientists already know how to answer. His problem with representations of DNA, genes, and the genome is what he calls the "hyperbolic excess of a vulgar understanding" that the "genes 'create us, body and mind.'"[6] Quoting Dawkins for this incident of excess, Lewontin points to the only book critical of these claims, Ruth Hubbard and Elijah Wald's *Exploding the Gene Myth.*[7]

Hubbard and Wald focus on the quality of the biological thinking that has accompanied the explosion of claims for DNA and genetic causality. In a sweeping survey of Mendelian genetics, issues of privacy, and DNA identification, Hubbard and Wald set out to provide a "realistic sense of the positive contributions genetics and biotechnology can make, and of the risks inherent in the science, its applications, and its commercialization."[8] Their book is an invaluable corrective that addresses and resituates claims about DNA and genes within a set of assumptions about health and disease and the science of genetics. *Exploding the Gene Myth* demythologizes claims about genetic causality, soberly pointing out, for example, how such claims as genetic bases for behaviors are often premised on cultural rather than scientific definitions of behavior.

Although Lewontin recommends Hubbard and Wald's book, he complains that "after the myths are exploded there is nothing left but a hole in the ground. The truth about alcoholism, violence, and divorces is that we don't know the truth."[9] Lewontin's symptomatic plaint itself reveals an attitude about science that underlies most of his critiques: that we should find a truth. The problem with such a "truth" is that it still depends on an Enlightenment notion of truth as an answer to a specific set of questions where cause and effect can be traced. Such a truth, even if it exists, can never escape the twistings and shadings of its representation, particularly in the public sector for a nonscientific audience.

If science itself is part of the problem with the excesses surrounding DNA and genes, then representation and ideology are another. Corrective and demythologizing texts such as those produced by Lewontin or Hubbard and Wald deploy a style of "plain speak" to counter what they see in their various ways as the excessive claims of science and science journalism. This plain speak tries assiduously to avoid employing analogies, and for the most part Hubbard and Wald succeed admirably at this. Lewontin, who is far more polemical in his intriguing debunkings of representations of biology, is both aware of the ideological situating of science and somewhat less equipped to deal with this ideology as anything other than ideology understood as a kind of vested perversion of the truth. For example, in his attack on the hyperbole surrounding the genome in "The Dream of the Human Genome," Lewontin cleverly introduces his essay by defining fetish and goes on to point out how figures such as the Holy Grail situate the genome inaccurately as the quest for the secret of everything. Lewontin is canny in pointing out how figurations of the genome make impossible claims for it and centralize this project to the exclusion of other worthy approaches; like Hubbard and Wald, his mission is to debunk these claims. But, in addition, Lewontin begins to analyze the ideological impetus of these figurations as misleading and wrong, a task that involves him in a kind of corrective thinking—if the Grail is a misleading figuration, what representation would not be?—that also entangles him in other figurations, in this case in a counterargument based on the metaphor of language. What he doesn't notice is how language itself is a misleading metaphor.

Lewontin tirelessly aims at isolating misleading representations, correcting them and even hypothesizing motives for their appearance within the funding politics of science. But Lewontin's task is not to analyze exactly how and why the metaphors and analogies appear as they do, a topic less about science than about the interplay of representation and ideology as it manifests in public discourse. This is the job of humanists.

Two other writers, however, do approach the problem of representing the gene. In *The Misunderstood Gene,* Michel Morange attempts to correct misperceptions about the gene, endeavoring to update the public with more recent understandings of genetic processes.[10] In his outline of current conceptions of the gene, Morange quickly draws attention to the dangers of metaphor: "This memorable image [Mendel's library] illustrates the importance of linguistic and computing metaphors in modern biology and, in particular, in the study of gene function. And it is the very power of the metaphors that are linked to the concept of the gene that strikes fear into the hearts of those who worry about the direction of genetic research."[11] Morange demonstrates, for example, how thinking of genes as a book "elevates genes above all other cellular components, which . . . is a misguided way of looking at the internal organization of organisms."[12] He also points out the weaknesses of the metaphor of the computer program: "It suggests that development can be directly decoded from the genes. At best, knowledge of genes can reveal the structure of the proteins that lead to the progressive construction of the organism through complex action in cells, tissues, and organs. The program metaphor also suggests that, in the organic world as in the world of computing, it should be possible to distinguish the software from the hardware."[13]

Morange's introductory chapter begins an important inquiry into how metaphor replaces one process with another. This, however, is not finally the project of Morange's book, which proceeds to provide a more accurate resituation of genetic processes. Morange dismisses the problem of the metaphor by simply blaming it on the carelessness of scientists: "Should we rid modern biology of its computing metaphors in order to avoid being misled? Some people think so. However, it is very unlikely that biologists, who have never been much bothered by impure, vague language, will consent to do this."[14] Scientists, however culpable or careless they may be, are not entirely responsible for the kinds of powers of the analogies they use to describe DNA and genes. Scientists exist within a larger history of ideas, assumptions, views, and what Jean-François Lyotard calls "legitimating metanarratives" that swirl around in what we might understand as a cultural unconscious. Scientists' choices are often less choices than available models attached to contemporaneous assumptions about how the world works.[15]

In two careful, convincing, and influential books about the relation between the idea of the gene and science's conceptual vocabularies, Evelyn Fox Keller addresses the problem of the larger environment of assumptions, or at least the environment of assumptions that drives scientific inquiry. In *Refiguring Life,* Keller asks, "But what does attributing (or for that matter denying) causal power to genes mean? To what extent does this way of talking reflect a set of 'natural facts,' and to what extent does it reflect the facts of a particular disciplinary culture? And is it just a way of talking? Is it not also a way of thinking, a way of seeing, and a way of doing a science?"[16] Her analysis of the problem of language in both *Refiguring Life* and her next book, *The Century of the Gene,* explores crucial shifts in the conceptual vocabularies we employ to understand scientific concepts as those occur primarily in the discourses of science itself.[17] Tracing, for example, the battle between the discourses of embryology and genetics, Keller shows how the notions of the nucleus and the cytoplasm become gendered—the nucleus masculine, the cytoplasm feminine, which accounts in part for "the discounting of maternal effects," a term she notes disappears even from indexes.[18] Informing her reading of the shift in our understandings of what information might mean and how it can work biologically, Keller situates Jacques Lacan's comments on science in tandem with physics, and in particular with the theoretical physicist Erwin Schrödinger's *What Is Life?,* showing how the figuration of the gene as an agent is linked to a modernist notion of the subject.[19]

Keller's work begins a crucial investigation into the intersections of language, ideology, the history of ideas, and science. In *The Century of the Gene,* she traces the evolution of the idea of the gene from its early-twentieth-century inception as well as the effects the concept has had on "biological thought."[20] Tracing transformations in the ways scientists have thought about genes and the genome, Keller shows how discourse has affected scientific practice, from the idea of the gene as a material basis for biology to the problem of the multivalence of the term *gene.* When the basic term for an immense set of practices and concepts begins to function as just that—a single term for many different processes—language becomes an impediment instead of a means of communication. Quoting the prescient Danish scientist Wilhelm Johannsen, who coined the term *gene,* Keller suggests that new terminology would help resolve the confusion, ambiguity, and ideological biases that have accrued around the gene: "It is a well established fact," Keller quotes Johannsen as saying, "that language is not only our servant, when we wish to express—or even to conceal—our thoughts, but that it may also be our master, overpowering us by means of the notions attached to the current words."[21]

Concerned with the question of "function"—the function of individual genes, the function of the entire genome—Keller demonstrates the complex interrelationship between the linguistic and conceptual connotations of the terms appended to genes and the genome and the kinds of assumptions scientists have employed in biological research that may well have been inflected by terminology. The last third of *The Century of the Gene* explores the import of the metaphor of the computer program in assumptions about the genome's functioning. The software metaphor has accumulated confusion, as it is "ever more entangled by the interweaving of ideas, skills, and vocabulary among their home disciplines and, perhaps more bewilderingly, by new modes of material construction and intervention."[22] Indeed, Keller declares that "it becomes difficult at times to know which is serving as a metaphor for the other, or even to distinguish our descriptions of one system from those of the other."[23]

Keller's invaluable intervention continues and sustains an ongoing inquiry into "popular images of the gene," taking on such basic and "deeply embedded" misapprehensions as the idea that genes are "clear and distinct causal agents, constituting the basis of all aspects of organismic life."[24] Usefully paralleling Keller's work is Richard Doyle's examination of the "rhetorical collision" between "the discourses of the life sciences" and "the discourse on life" in Schrödinger's *What Is Life?*[25] Doyle shows how in Schrödinger's work, the notion of the pattern or phenotype of an organism becomes contained within the chromosome as a genotype, rhetorically turning "'pattern' and the 'organism' inside out. . . . injecting the life of the organism into its description."[26] The concept of the "pattern," thus, enacts a shift from a complex observation of the literal to a figuration that displaces both process and organism into a trope. Thus "mistaking a text for complete development or concept," Schrödinger metaphorizes, synecdochizes, and, more important, conceptually inverts the complexities of the interrelations of DNA and organisms, reproducing in the chromosome the model of the brain or library where control is located and by which agency is effected.[27]

Doyle's sophisticated work traces the "rhetorical transformations" and "slippages" by which life as the object of biology becomes reduced to the molecule. Other scholars such as Dorothy Nelkin and M. Susan Lindee, José van Dijck, and Celeste Condit have all explored the rhetoric surrounding the public discussion of genes and the genome. In *The DNA Mystique,* Nelkin and Lindee address the effects and powers of representations of genes in popular culture. As sociologists, they take such representations as "folklore" appearing across many popular discourses, which "both reflect and affect the cultural ethos," and in which the figure of the gene serves needs by promoting products, helping groups "proclaim their solidarity," explaining success or failure, or supporting positions in relation to various causes.[28] Specifically, Nelkin and Lindee identify and address such genetic themes as signifying identity and explaining disease, immortality, human relatedness, good or bad behaviors, and as mixing into issues of race, gender, and sexual orientation. Their relatively early study produces a fairly thorough map of the major themes and ideas that genes represent in contemporary culture.

Van Dijck continues Nelkin and Lindee's work in *Imagenation: Popular Images of Genetics,* elaborating on the "role" that "images and imagination" play "in popular representations of the new genetics since the late 1950s."[29] Focusing specifically on human genetics in popular (as opposed to scientific or professional) nonfictional representations, van Dijck shows how genetics functions as a site of contestation as well as how genetic science itself has deployed what she terms an "imagenation" to shift less acceptable paradigms concerned with morality and safety into more exciting and profitable paradigms such as "mapping." The book then explains "the recurring paradox between changing technologies and persistent representations thereof" as a way to understand the play of ideologies and tropes of acceptance.[30]

Taking up Nelkin and Lindee's sociological approach, Condit offers a "quantitative and critical interpretation" of "the range of public meanings about genetics."[31] Condit's *Meanings of the Gene* explores issues of

that make DNA seem far more transparent, malleable, accessible, and ownable than it is. While it may not be unreasonable to think of the genome as a book, such a comparison is plausible because we already think of DNA as an alphabet. The question is why we think of it as an alphabet in the first place. Surely that analogy does not derive simply from biologists' convention of referring to each of DNA's four nucleic acids—adenine, guanine, cytosine, and thymine—by their first letters. If that were the case, all of chemistry would be an alphabet, all molecules sentences, and the entire universe something like the *Encyclopaedia Britannica*.

What have analogies such as the alphabet, the book, the recipe, or the homunculus contributed to the conceptual mix of science and culture in the late twentieth century? The kinds of analogies employed and the processes and values swimming in the admixture point to a whole way of thinking about life, science, knowledge, property, and the individual in the past fifty years. This way of thinking neither precedes nor devolves from science but instead exists as a part of a complex cultural system that positions "science" itself as a particular kind of "truth." An equally important part of this system is how such cultural values as gender or capitalism emerge as scientific "truths," the latter at least insofar as its processes are elicited as "natural" models for competitive dynamics (and vice versa—as the model for competitions in nature). At issue in all of this is not the "truth"-value of DNA or the fine complexities of molecular biology but how scientific artifacts such as DNA function simultaneously as cultural icons and vectors for cultural and ideological work.

The Narrative DNA Gene: A Pseudoscience We Can't Take Seriously

White-coated geeks wearing thick glasses hunch over laboratory benches crowded with glass. Rainbows of bubbling liquids follow mazes of tubes, pipes, coils, and beakers. Bespectacled men peer intently, adding a few drops of this, a few drops of that. They stir. They breathe deeply, square their shoulders, and drink the elixirs they have distilled. Change comes slowly, then violently. They thrash through jungles of equipment, shedding coats and glasses, sprouting hair and muscle and an entire new wardrobe of facial expressions. They become debonair.

This is a typical scenario in film comedies about laboratory science. In comedies that parallel the arc of DNA's emergence—*Monkey Business* (1952), *The Nutty Professor* (1963), *The Nutty Professor* (1996), and *Nutty Professor II: The Klumps* (2000)—goofy scientist characters attain normalcy (or supernormalcy) by ingesting a "formula." Comic potions from films and cartoons (remember Popeye's spinach?) have the power to alter physique and personality in much the same way that Dr. Jekyll's more calamitous experiment transformed him into Mr. Hyde. But the formulas in film comedies work in a more fortunate direction, transforming nice but socially inept scientists into irresistibly attractive Romeos. Unfortunately, the formulas lack staying power, and the laboratory lotharios revert inopportunely to their former selves.

Although film comedy might be only a frivolous reflection of the sustained and serious endeavors of scientific research in the twentieth century, it nonetheless exposes and explores beliefs, ideas, and anxieties about science that circulate more widely in culture. Unlike science fiction—even in its comic versions, which as a genre tries to provide at least the hint of a scientific explanation for its miracles—science comedies take over and amplify cultural beliefs as a part of their generic working. Popular misconceptions about science become part of the stuff of the comedy so that comedies are much more symptomatic readings of myth than more "serious," or even fantasy, genres might be. Comic films about the transformative power of formulas provide insight into the fantasy that will come to be occupied by DNA, the molecular arrangement that replaces mysterious and often accidental concoctions as a powerful transforming agent. Like the film scientists' compounds, DNA is imagined as a magic potion that offers possibilities for transformation. That film comedies shift from the colorful chemical formulas of the 1950s and 1960s to the computer-imaged DNA of the 1990s as the symbol of the powers of laboratory science mirrors the accruing powers of DNA as well as its increasing representation as a visible, accessible, legible agent of development and change. The appearance of DNA as a malleable laboratory tool in 1990s science comedies presents both dream and anxieties about DNA's power to alter the fabric of life as well as to engage the pseudoscientific magic that will become DNA's constant companion.

The power of DNA as it appears in film comedy is a narrative power, one of the most basic, pervasive, and least-acknowledged processes in contemporary culture. Narrative—the shape of a story—is a sense-making tool, a way to arrange cause and effect to produce a linear, sensible scenario. Narrative brings the assumption of certain values and cause-effect relationships with it. We assume, for example, that conflict will eventually produce something like a happy ending, knowledge, a victory, a product, a marriage, a child. This notion of production parallels our ideas of capitalist investment and payoff as well as our imagination of a heterosexual reproductive scenario.[32] In this environment, science itself is transformed from a complex set of processes, systems, and stages to a matter of instant and naturalized causality. In addition, narrativizing scientific phenomena recasts these phenomena as gendered domestic comedies (or tragedies) about transformation itself. How many times, for example, have we witnessed renditions of human conception as epic romance between a stalwart sperm fighting against the odds to reach a bumbling, passive egg?

As narratives about transformation and its failures, laboratory comedies uncannily play out a cultural transformation from the chemical formula (as the often colorful substitute for the pharmaceutical) to DNA as an agent of immediate and unmediated transformation. These screwy lab comedies imagine the pros and cons of a science that works like a story in a context driven by considerations of profit and romance. They envision science as the discovery of a single formula that alters the course of personal lives and human development and evolution in general. *Monkey Business,* released a year before Watson and Crick announced their discovery of DNA's structure, follows the antics of a corporate research scientist who accidentally discovers a formula for the fountain of youth. Directed by Howard Hawks (who was renowned for his screwball comedies) and starring Cary Grant as a seriously myopic and absent-minded scientist, Ginger Rogers as his commonsense wife, and a young Marilyn Monroe as the boss's secretary, the film prophetically combines three of the elements that will come to characterize popular depictions of DNA: (1) the centrality of a keylike formula that intervenes in life processes; (2) an intermixing or confusion of species that suggests a tampering with evolution; and (3) the motive of corporate profit.

The film centers on a formula concocted by a mischievous lab chimp who pours it into the laboratory drinking water. This formula throws its imbibers into reverse development. Grant drinks from the water cooler, sheds his glasses, buys a new wardrobe and a sports car, and squires the secretary around town. His wife then drinks the formula as well and becomes childlike. The rejuvenation processes ultimately put the unwitting subjects to sleep, after which they awaken as their former "old" selves. Finally, when Grant is missing and the chimp appears in his place, his wife believes he has rejuvenated himself back into an earlier species. As we might expect, the president of the corporation for which the scientist works wants to market the formula, but a recalcitrant Grant, taking refuge in a put-on performance of childish obstreperousness, refuses. The story follows a series of predictable mix-ups and mishaps righted when Grant discovers the formula's source and disposes of it.

Eleven years later, another geeky scientist, this one a college professor, experiments with a similar formula. As a chemistry teacher prone to explosions, *The Nutty Professor*'s Jerry Lewis falls in love with a winsome student and works to concoct a formula that will transform him from the nearsighted, bucktoothed, screeching Professor Kelp to a suave, magnetic, baritone playboy called Buddy Love. The potion works, the professor (as Buddy Love) manages to woo the student, but the effects wear off too soon, forcing the cracking-voiced professor to flee mysteriously. The professor's nemesis is the dean, a vain, controlling, bottom-line administrator who is anxious to preserve his own control and who easily dominates the timid professor. Enlisted to help entertain at the senior prom, Professor Kelp as Buddy finally runs out of luck, losing his playboy persona in the middle of the performance. He confesses his excess and pronounces the lesson he has learned: it is better to leave nature alone, aspirations to coolness will backfire, and nerds are really lovable for themselves.

Despite his humility, in a moment of meditation before his final performance Professor Kelp recognizes that the persona he becomes when he is Buddy Love is not entirely produced by the chemical he takes. "Heredity," he exclaims. The formula somehow taps into the oedipal drama of his childhood, which a flashback shows us is a spectacular case of a domineering wife and henpecked husband. The personality that the chemicals release has always been a hidden (recessive) part of the professor's disposition as his response to his mother's humiliation of his father. The formula taps into heredity *and* psychology, physique and behavior, not unlike the powers we have recently ascribed to genes.

In 1996 Eddie Murphy revives Lewis's professor in a remake of *The Nutty Professor* and its 2000 sequel, *Nutty Professor II: The Klumps*. These two films, which share similar plots, combine elements of *Monkey Business* with features of the original *Nutty Professor*. Professor Klump, engaged in performance enhancement experiments with hamsters, has discovered a way to improve their growth and strength with a potion that uses DNA to augment the hamsters' genes. The chemical formula of the 1950s and 1960s has been remade into a new and better DNA version, which acts more or less the same as the formula. Falling in love with a young, beautiful colleague, the professor, an obese but kindly nice guy, desperately takes the potion, transforming himself into an obnoxious, charismatic, hyperactive macho man. The college dean, a mixture of *Monkey Business*'s mogul and *Nutty Professor*'s dubious administrator, wants to use Klump's formula to attract a particularly generous donor. Going further than his comic precursors, Professor Klump regresses to the point of literal infancy as Buddy Love, his fountain of youth formula working too well. The Buddy Love side also experiences interspecies mixing, as some canine DNA has gotten mixed up with his DNA magic bullet. As in the earlier comedies, this latest nutty professor sees the light and learns that he lives better without chemistry.

The developmental narratives that structure science comedies are simply versions of the basic pattern of narrative in Western culture, which is itself another kind of "formula." The omnipresence of this same basic story in conceptions of most kinds of biological, historical, political, and economic processes links them to one another as well as to the same general pattern—to the same old story. The story's end connects success, (re)productivity, and fulfillment (individual, social, and historical). The life story becomes the story of capitalist production in which capital is invested, grows, matures, and pays off. Advances in human culture are reflected in material being created by capitalism—or what we understand as "progress." The goals of one kind of story—the story of capitalist investment—might substitute for the goals of another story: successful human reproduction, scientific research, or cultural and social progress. In this way investment might spur reproduction, reproduction might effect knowledge, knowledge might result in a big payoff, and humanity is in a better position than before.

Instead of reflecting what actually might happen in, say, molecular biology, these popular renditions of science present biological phenomena through the same formulaic narrative that pervades Western culture. Comedies employ rejuvenation narratives that alter the middle of this story as a way to put off or change the end. Comic science films feature the individual version of this story, which is also, as I shall show, the basic narrative of disease. In these an individual is born but grows up as too fat, ugly, or geeky and is thus destined for an unhappy end. Chemical formulas intercede to alter the individual's detour into idiosyncrasy, making (usually) his story normal—that is, successful and fulfilling, especially in the area of romance. The problem is that magic bullets introduce chaos and disappointment that teach us that the normal course of events was okay in the first place. In intervening in the story, science is the wrong story.

The pervasiveness of this basic narrative paradigm means that it is common to represent the problems science addresses as flawed versions of the story. Scientific problems are inevitably linked both to potential changes in the plot (i.e., a better middle means a satisfying end) and to capitalism as the necessary economic environment of scientific achievement. In a much larger arena, the commonality of narrative patterns among accounts of life, biological processes, and history makes it easy to link changes in individual narratives to potential interventions or disturbances in the story of evolution, as evolution itself is transposed into a narrative of progress through hierarchies of species' complexity and importance. To link science, lives, and capital to depict scientific endeavor does not mean that there isn't also some real-life connection among them.

However, the narrative connection naturalizes their relation, making science, lives, history, and capital seem inevitably intertwined and intrinsic elements of one another. Thus science always exists within this "story," and the story is both its most persuasive rhetoric and the means by which science is no longer science at all but a kind of fiction. That scientific formulas are ultimately unsuccessful at changing the story (even as they produce a story or even promise change) suggests the stubbornness of that basic story even in the face of a rapidly changing innovative science. Representing science as a story means that very little actual science is promulgated at all.

The Magic Gene and the Miracles of Pseudoscience

The humanities might be able to show science (if indeed these are two separate realms) how the relation between science and representation produces a paradox that is self-contained in the figure of DNA's double helix. On the one hand, representations of scientific fact are always more than fact, importing values and ways of thinking with words and analogies. The specific figures representing scientific facts situate such facts as a part of a complex system of cultural values in which the "fact" functions as a fact among others (and hence not the only "truth" or even "truth" at all, if we understand truth to be singular). We might, for example, accept evolution as fact or reasoned theory and hence a species of truth. The theory of evolution has become linked popularly to images of both chimpanzees and fish with legs crawling out of some primordial sea. Contemporary proponents of evolution have borrowed this ichthyic iconography of Christianity, which in its more fundamentalist form adheres to alternate and opposing ideas of creationism or intelligent design. The fish figure in its various forms (with legs or without) plays out the war of basic beliefs and alternative truths in a culture in which science cannot persist outside a system of cultural values. The coexistence of two theories each claiming truth elicits either a denial of one or the other, or produces a sense that multiple kinds of truth might coexist, making truth itself relative rather than absolute. It also serves as a means by which individuals can position themselves in a larger matrix of competing values, as competing fish outlines, with or without feet, are displayed on the backs of automobiles.

The other side of the culture-science paradox is that the ideas that analogy, metaphor, and narrative add to the mix sometimes become themselves the transposed truths of science. We might call the genome the "book of life" as a convenient analogy, but the idea of the book persists in the ways we understand the powers and mechanisms of DNA. The human genome becomes a language that genetic medicine hopes to rewrite—an analogy that incarnates both a continuation of the book figure and an operative literalization that has been used to conceptualize gene therapies. Understanding genes as agents (even agents with selfish motives) makes it easier to think of genes as a strategic site to focus palliative measures.

The paradox—that representations of science render scientific facts less "true" (or more culturally relative) while the figures of their representation become scientifically operative—is a paradox only within the larger realm of cultural dynamics. Although we might question the status of any fact as always provisional, the effects of modes of representing scientific facts and theories add obvious sets of additional material that inevitably translate what we might understand as objective phenomena into cultural artifacts—if such ideas as a "fact" or "theory" weren't already artifacts to begin with. The problem is that context is unavoidable, so that there is no pure science to be had, even though we may believe such a thing exists somewhere. Although the status of science as truth is a different problem than the importation of value into the figurative representation of scientific phenomena, the two are related insofar as they both share in the larger systems of cultural value and ways of thinking, reflecting different aspects of the same system.

Looking at the ways DNA is represented may help unfurl the paradox of science and its representations, making us conscious of how something as insignificant as signification can unravel the science of science. While no poststructuralist would ever believe in the exclusive truth-value of science—science is itself always a part of a complex cultural system—it is well worth the effort to see how the representations of science become the science of representation. Representations of science play out entire cultural dramas, since apparently the truth that science threatens must be transformed into values that feed the cultural system, even as those facts might ultimately transform it.

DNA, however, is not just another scientific fact. DNA's overt connection to processes of representation (the alphabet, the book, the map) makes the representations of DNA particularly rich sites for understanding the interrelation of science, metaphor, and narrative. Figured as a life language and special case, DNA works as the nexus between life, science, the mechanics of expression, and reproduction, redoubling the relationship between science and representation and forging a connection between representation (or standing for) and reproduction (making more of the same). DNA "represents" (its "letters" "spell" genes) in a way imagined to be analogous to how we try to represent or convey DNA's structure, functions, and operations. While DNA representations reproduce more DNA, our representations reproduce it along with the ideas and biases of our culture. DNA's representational capabilities—its functions as book, language, code, map, blueprint—are themselves the subject (and partly the products) of representation, creating a kind of self-referential circle in which the figuration of DNA's representational functions refers to both biological and cultural reproductions of the same. Like its own double helix, DNA both doubles and reproduces representations, making it a particularly productive site for understanding the interplay of cultural processes and ideologies that accompany epistemological and systemic change. Figuring the agency through which we imagine that bodies are produced and changed, DNA operates as both causal and masking agent, betokening simultaneously science and myth, perpetuation and transformation, the molecular and the gross, fear and salvation.

The Compensatory Functions of Representation

The figure of DNA—and the DNA gene—obscures and enables processes of deep change in the symbolic bases of culture and its logics of relation and organization. The way DNA and DNA genes actually work represents a significant change in how we have traditionally conceived of life. Such events as the production of individual identity, reproduction, family, and death are instances of coherent patterns. Such practices as naming, symbolic (or homeopathic) substitutions, kinship structures, and myth or storytelling have provided conceptual order. In this system of metaphor (the process of substitution and symbolization) and narrative (the paradigm of cause and effect ordering), we name things, which substitutes a word for a thing. Naming symbolically places individuals within the kinship structures of culture, and the kinship structure provides the names. We repeat patriarchal family structures as the truth of religions. We have imagined lives as following a particular trajectory or story with a beginning, a middle, and an end; morning, noon, and night; spring, summer, fall, winter. We envision cells and atoms as little cosmos, and we have lived in a cosmos where words and rituals have defined and given meaning to birth, puberty, marriage, and death. We have extended our imaginings throughout this cosmos and have claimed it as the realm of our gods.

We have, in other words, lived in a universe made sensible through patterns repeated on every scale, defined by the words we use to name it and the stories and metaphors we use to explain it and that come to substitute for it. As the mechanism by which a word stands for the thing to which it refers, metaphor is also a logic of comparison in which a series of relations is produced through analogy, but in which there is no necessary connection between the ideas compared. This logic of metaphor represents a leap between the thing or phenomenon named and the thing itself, representing a kind of magical sense-making apparatus in which a few analogies come to define a large number of unrelated and essentially dissimilar events and processes. For example, there is at best only a rough correlation between our narratives of the shapes of lives and how human organisms survive, yet we understand the lives of humans and the lives of paramecia within the same paradigm: birth, reproduction, death. We apply hierarchies to species that are like the hierarchies of human organizations. We think of the earth revolving around the sun in the same way that cell matter surrounds a nucleus or electrons circle an atom's nucleus. The repetition of kinship patterns as defining cosmic events in creation myths or our earlier understandings of human bodies as composed of four symbolic humors has centered human beings as both the pattern for and instances of the patterns of a divine order. And this divine is the grandest metaphor of all, representing the course of human wishes and projections.

Following other inventions from advanced industrialization, DNA, however, does not literally function through any leap of metaphor, symbolization, or even life pattern. Rather, like the geared machine whose sprockets must properly interlace, the nucleic acids of DNA function through contact with one another. Substitutions are errors that often cause malfunctions. DNA's logic is a logic of contiguity (or metonymy)—of how each site of intersection affects each other site. In addition, instead of the repetition of the same pattern through different levels or scales, a logic of contiguity represents a logic of systems whose possibilities go multiple directions at once. The contiguity represented by geared machines, DNA, and, most recently, digital computers (whose calculations are accomplished through a chained series of linked circuits) is a logic at variance with the practices of substitution and symbolization we have relied on for so long. Our tendency is to relegate machines and computers to the realm of the mechanical, a service arena that has only a subordinate relationship to larger patterns of significance. Hence the machine is subordinated to its task, and its workings often masked to give us the illusions of instantaneity and metaphor. A toaster toasts without us understanding how its coils use electricity. An automobile runs, more or less on the analogy of the human body, which eats fuel and expels wastes. A computer processes, masked by metaphoric interfaces such as Microsoft Windows, designed to make us think we are picking tasks and items at will, like shopping.

Translating the operations of the mechanical into metaphors and analogies masks them, keeping in place notions of commodity, utility, hierarchy, and order typical of our metaphoric symbolic structures. These are, after all, machines. But what happens when something mechanical—like DNA—becomes the pretext for life? What happens when computers and computer logics begin to illustrate and substantiate other kinds of logic that are ordered differently from metaphor?

The DNA gene is the specimen that begins to make this masking visible. Computers retain—and in fact increasingly mask—their processes.

We take machines for granted, as automobile engines are increasingly enclosed from view and computerized. We buy appliances (pace the visible dirt collector on some vacuum cleaners) where complex processes normally undertaken by humans in collaboration with several machines occur within a gleaming black-and-silver facade. (Some coffeemakers now grind beans, brew coffee, and expel the grounds automatically and invisibly.) We no longer see the mercury column on a thermometer, food is mysteriously zapped in microwaves, minute hard drives spew music from gum-package-sized players, heating elements are enclosed behind a layer of ceramic—in short, the mechanical processes of living are increasingly hidden from view. At the same time, the DNA gene with all of its capabilities seeds the media, providing proof of human ingenuity (we are our genes) while implying that such ingenuities are the accidents of biochemistry, environment, and chance. Strategies for representing DNA to the public focus on human ingenuity. Science focuses on the rest.

This book examines how modes of identifying and describing the DNA gene function in contemporary culture to allay fears about changes in order and the logics of systems and to rewrite the truth of humanity in safe and conservative terms. There are two basic arguments. First, the ways we think of DNA and genes are themselves the logical product of centuries of thought. DNA isn't what it is because that is what it is. Rather, the emphasis on structure and function attached to our understandings of DNA are the capping response to several centuries of reductionism and dialecticism. Second, the analogies derived from DNA's position as a structural answer to questions of life and deployed to describe and explain DNA genes are also metaphors that import particular compensations or remedies for the cultural fears excited by the discovery of DNA and other systemic ways of thinking. The ways we conceived of DNA and genes in the second half of the twentieth century are not only an effect of a history of thought that ends up with the idea of structure as an answer in itself, they also perpetuated this notion of structure at the very moment they imported alternative ideas of system and complexity.

The following chapters trace and analyze the sets of ideas that have come into play in attempts to present DNA genes to the general public, demonstrating how apparently simple analogies convey complex sets of ideas that respond to contemporary anxieties and interests. The second chapter, "Genesis," traces the conceptual family tree of the DNA gene, showing how the gene is the "natural" heir to a mid-twentieth-century convergence of structuralism and reductionism. Following several trails of thought from the Greek philosophers to the more recent inventors of cybernetics, psychoanalysis, and systems theory, the second chapter suggests that if there hadn't been such a thing as a DNA gene, we would have contrived it anyway, since the DNA gene is the point at which many long-lived ideas about the order of the universe converge. It argues that our conceptions of the DNA gene as the secret of life are already conditioned by our ideas about language and binary modes of organizing knowledge. It also argues that the forms our understandings of DNA take are themselves already the defensive and compensatory adoption of the more familiar forms of structuralism such as a code or language in the face of the more threatening epistemologies of the equally contemporaneous (but much less conventional) systems theory, which might have provided a more accurate and less exploitable set of genetic concepts.

Chapter 3, "Flesh Made Word," examines the uses and effects of textual metaphors such as the book of life, the code, the blueprint, alphabet, or recipe employed to describe DNA, suggesting that these textual metaphors produce a continued sense of human control and agency over genetic processes and provide the conceptual basis for turning genes into property via patents. They also enable structural fantasies that override far more complex ideas of system, complexity, chaos, and other ways to understand the interrelation of phenomena.

Chapter 4, "The Homunculus and Saturating Tales," investigates how the metaphor of the gene as the homunculus, or little man, combines with the invasion of narrative logic to produce an idea of the gene that perpetuates conventional ideas about gender, sexuality, and delusions of immortality. Looking at the relations between press announcements about genetic "discoveries" (e.g., the "gay" gene) and popular interpretations about the meaning of genetic capabilities, this chapter explores how familial ideologies are reinstituted and rendered scientific "truth" via the gene.

The fifth chapter, "The Ecstasies of Pseudoscience," focuses on how the science of DNA is constantly rewritten as a version of pseudoscientific magic, from the fantasies of genetic tinkering presented in movies to discussions of "intelligent design" to the more strategic imaginations of gene therapies and engineering. It also examines the problem of how any representation of science could be anything other than a misrepresentation.

The final chapter, "Rewriting History," examines the transposition of DNA into individual identity as well as a commodity where one can, for example, put one's genome up for bid on eBay or define oneself in relation to a DNA profile purchased over the Internet. This chapter investigates the structural relation between DNA imagined as an information product, contemporary pressures on an essentialized identity, and how both DNA information and identities reflect transnational commodity culture. The DNA commodity both masks and reveals the evolution of the postcapitalist commodity system from one of the phantasmic exchange of commodities to an economy of perpetuated subscription and debt masquerading as an access-to-information product. Subtended by DNA's code analogy, this new economy pays it forward, so to speak, in that it provokes a present system of payment in exchange for the hoped-for future value of information and uses the code may generate. This is disturbingly analogous to contemporary perpetuated credit systems such as car leases, credit cards, utilities, and other subscription schemes that indenture the consumer in a perpetual web of payment for a commodity that no longer has a location.

DNA genes represent one of the fundamental advances of twentieth-century science. While medical possibilities are one benefit we might enjoy, another—and perhaps ultimately more revolutionary possibility—exists in the kind of systemic organization DNA represents. When we mask, translate, or otherwise obscure the notion of system, we cling to the past, allay our anxieties about the future, and slow down changes in how we might view the centrality of the human, the permanence of our lives, and the evolving survival of the biosphere.

the word. Not only do we see how life and matter come together in the same chain, we also understand the individuality of organisms—their identity synecdochized by an idiosyncratic ordering of nucleotides.

As simultaneously conclusion and solution—as a kind of universal synthesis—DNA occupies two broad but contradictory positions in relation to the history of ideas. As the premier example of structure equaling function, it embodies the perfect structuralist system at the peak of structuralist fervor in the 1950s, itself the heir to Enlightenment logics and nineteenth-century dialecticism. Yet precisely because DNA is probably the most famous example of a self-identical functional structure regarded as almost infinitely meaningful, it masks a shift to the contemporaneous emergence of less structuralist, less dialectical (or more poststructuralist) ways of thinking about phenomena. Looking like the ideal of performative structural complementarity, DNA's "decodings" are multiple, variable, and contextual—more complex and less strictly dialectical than popular renditions would have it. In other words, DNA also bears the impress of emerging nonlinear modes of analysis involving complex, multidimensional, nonbinary logics introduced by theories of relativity (Einstein), particle physics (Richard Feynman), systems theories (Norbert Wiener), and poststructuralist thinking of all kinds.[1]

Part of DNA's ability to distract us from, obscure, and perhaps signal threatening new ideas comes from the fact that we focus on DNA as the simple yet efficacious key to the operations of a much more complex system. Our relative comfort with the concept of DNA comes through analogy—though the fact is that we already imagine the DNA gene sharing in the same dialectical structures that we use to understand everything from language to kinship to law to history. For this reason, such potentially order-shaking possibilities as cloning, genetic engineering, and gene-based cures, though vaguely threatening, seem already familiar as simple versions of textual manipulation (copying, correcting, and amending). But there is also an irony: while our idea of DNA centers the human as the agent of knowledge and the discoverer and decoder of a code that unfolds an orderly structure, DNA's pervasiveness also threatens to decenter humanity as either central or biologically special, making people mere vessels for the perpetuation of a chemical that is a common denominator among species through history.

Particle-arity, or Method Becomes Material

How and why have our quests for knowledge arrived finally at the DNA gene as the prime mover, the molecular cause of all biological and many behavioral phenomena? Why has the increasingly microscopic and infinitesimal been seen as the site of fundamental processes? Small, primary elements represent a locus of knowledge—of having gotten to the bottom of things. A particle that is common to everything (or at least to all life); that behaves in a consistent, orderly, stable, and predictable manner; and that provides a mechanism for its own perpetuation satisfies our notion of what could be foundational and true. If we can locate these smallest things, then we can find out how they combine to produce larger phenomena. The result is a form or structure that ultimately defines how something works. Although all evidence might suggest that minute particles of some kind do indeed ground the workings of the physical universe, how we arrived at the view that minute particles, whether atoms, quarks, strings, or genes, work as the elemental organizational elements of matter and life—indeed, how we arrived at the idea of the particle as elemental—is also an effect of the ways we think about phenomena, ways that haven't really changed for 2,500 years.

DNA is the promise of a long legacy, the coming together of two related tendencies—the reductionist analytic atomism of "hard" science and the structuralism of philosophy, linguistics, narrative, and cultural criticism. Reductionism is an inherent assumption of a mechanistic or analytic view of the world. Begun by the Greeks, analysis understands large phenomena as effects of the actions of many smaller, interconnected elements. This reductionist way of thinking pushes toward finding the smallest, most basic, primary, and indivisible element as the end (and beginning) point for all processes. The minute is also the ultimate answer to questions about origins, systems, and the operation of universal phenomena, including the origins and essence of life itself. In the fifth century B.C.E., Anaxagoras, Leucippus, and Democritus each formulated some notion of a basic particle, dubbed "seed" by Anaxagoras and "atom" by Democritus. Both believed that all matter was made up of unchangeable particles in a vacuum. The idea of the basic unchangeable particle lost out, however, to Empedocles' (fifth century B.C.E.) notion of four primary substances (earth, air, fire, water), taken up by both Plato and

Aristotle (fifth and fourth centuries B.C.E., respectively). This idea provided a much more graspable adjectival basis for a recipe (one that uncannily parallels the four complementary nucleic acids of DNA). Atomism returned in the seventeenth century in the ideas of Pierre Gasset (1592–1655), whose work just precedes the revolutionary insights of Sir Isaac Newton (1642–1727), and Robert Boyle (1627–91), who believed that gases were made of small particles. Biologists also looked for smaller structures. Marcello Malpighi (1628–94) discovered capillaries, Robert Hooke (1635–1703) named the cell, and a series of Dutch opticians invented the microscope in the early seventeenth century.[2]

By the early eighteenth century, Newton articulated an atomic theory consonant with a monotheistic universe. In *Optics,* Newton proposes a universe composed of tiny solid particles:

> All these things being considered, it seems probable to me, that God in the beginning formed matter in *solid, massy, hard, impenetrable, moveable particles,* of such sizes and figures, and with such other properties, and in such proportion to space, as most conduced to the end for which he formed them; and that these primitive particles, being solids, are incomparably harder than any porous bodies compounded of them; even so very hard, as never to wear or break in pieces; no ordinary power being able to divide what God himself made one in the first creation. (emphasis mine)[3]

Newton's atomism is one logical result of the assumptions employed by Greek philosophers to understand how the universe works, even if some of those philosophers, such as Aristotle, rejected the idea of the atom itself. Aristotle nonetheless still looked to the smallest basic unit as the location of answers for questions about how things work. For example, in thinking about both rhetoric and science, Aristotle proposed that "primary substances are most properly called substances in virtue of the fact that they are the entities which underlie everything else, and that everything else is either predicated of them or present in them."[4] In his considerations of science, Aristotle conceived of a system founded on primary substances and "original causes" that were themselves irreducible. "But everywhere," he says, "science deals chiefly with that which is primary and on which other things depend, and in virtue of which they get their names."[5]

Science, knowledge, truth exist in the primary and irreducible. Thinking about truth is a process of finding the "primary," "irreducible," "basic," "original," "underlying," "first," "substances," "principles," and "causes." Although there is nothing intrinsic to Aristotle's formulations that suggests that such primary causes need to be smaller than the phenomena they produce (and sometimes they are ideas instead of substances), the very notion of breaking something down into more basic elements—analysis—depends on a metaphor of spatial fragmentation, which, if taken literally, results in the impression that primary elements are smaller than the phenomena they underlie. In a similar way, primary causes or substances, because they are seen as primary or first, appear to precede more complex phenomena historically. We understand the world working from the primary to the complex, primary elements combining to produce more complex phenomena. Popular notions of the origins of life on the planet (even theories of alien "seeding") and conceptions of evolution exemplify this, as we understand life to begin from organic molecules and increase in size and complexity.

The idea that small primary substances underlie and explain phenomena provides the basis for both inductive and deductive modes of reasoning that underwrite the analytic scientific method as it developed in the seventeenth century. Whether scientists collect data and form a hypothesis (induction) or form a hypothesis and then test various irreducible elements predicted by such a hypothesis (deduction), both modes depend on the assumption that complex phenomena can be broken into simpler, more basic causes and elements, which, in a literalist logic, are the smallest of all.

The idea of analysis, or breaking into smaller and more basic units and phenomena, no longer represents simply a concept of the structure of the physical universe or an approach to the mystery of life. It also begins to incorporate the very terms and assumptions about what might constitute both reason and the truth. The object of analytic researches, and the methods and assumptions used to conduct such research, become one and the same. This Enlightenment merging of object and method anticipates, enables, and predicts our figuration of DNA as itself the conjoining of assumption, method, and matter.

The emergence of an analytic experimental method in the seventeenth century accompanied the reemergence of atomic theories that reduced problems to their most basic elements. By breaking into its smallest components the question of whether flies generated spontaneously from rotting meat—meat exposed to flies, meat sealed away from flies—Francesco Redi (1626–97) proved that flies did not spontaneously generate from rotting meat. Breaking a problem down implied a mechanistic view of living things. Phenomena could be divided into simpler, more basic processes whose combination would provide a mechanical, fact-based answer to how various life processes worked. At the same time, the discovery of the most basic components ratified general theories about how things work. The small is the proof of the large, a tendency we see today in the forensic fantasies associated with DNA.

As seventeenth-century scientists such as Newton theorized the rational complementarity of small and large, element and principle, technology enabled explorations to become both more distant (as in astronomy) and more minute.[6] Instruments that augmented vision such as the telescope and its inverse, the microscope, expanded the visible universe to scales unseen, which provided another theater for investigation. The capacity to see structures invisible to the naked eye was enabled by improvements in optics begun by Dutch lens makers around 1600. Hans Janssen (1590–?) and his son produced a microscope that permitted a detailed view of tiny insects. The microscope also provided the possibility that a more atomistic and mechanical view of life could be substantiated by finding basic, microscopic structures. Malpighi, who was influenced by Redi's methods, was drawn to a mechanical account of living processes. He employed the microscope to describe red blood cells and discern capillaries, thus filling in the gap in the theory of circulation developed by William Harvey (1578–1657). Malpighi's discovery and description of such microscopic structures as nerve fibers, alveoli in the lungs, and taste buds led him to envision the body as a mechanical phenomenon:

> The mechanisms of our bodies are composed of strings, thread, beams, levers, cloth, flowing fluids, cisterns, ducts, filters, sieves, and other similar mechanisms. Through studying these parts with the help of Anatomy, Philosophy and Mechanics, man has discovered their structure and function.[7]

Hooke, who first named the minute structures of wood "cells," also employed an experimental method that focused on "the real, the mechanical, the experimental Philosophy."[8] Hooke's method, as Porter suggests, subscribes to the revisions of the scientific method by Francis Bacon (1561–1626) in which observation of real phenomena replaces logic and philosophy and conclusions are drawn by an inductive method that begins with the "forms," or physical properties, of "causes." Forms, according to Bacon, will tell us the nature of the phenomenon. Form, or structure, derived from "particulars," revealing and aligned with axioms and principles, leads to their joinder in the intermediate:

> 104. . . . But we can then only augur well for the sciences, when the ascent shall proceed by a true scale and successive steps, without interruption or breach, from particulars to the lesser axioms, thence to the intermediate, (rising one above the other,) and lastly to the most general. For the lowest axioms differ but little from bare experiment, the highest and most general (as they are esteemed at present) are notional, abstract, and of no real weight. The intermediate are true, solid, full of life, and upon them depend the business and fortune of mankind; beyond these are the really general, but not abstract, axioms, which are truly limited by the intermediate.[9]

DNA is the "intermediate," the coming together, in Bacon's terms of the particular and the general—a mediator at whose point both analysis and principle come to life.

The search for small basic forms accompanies researches into properties and dynamics, though it is with reluctance that scientists abandon such concepts as "ether" and "miasma" (both environmental hypotheses) in favor of fields, germs, and molecules. In the mid-nineteenth century, Charles Darwin's theory of evolution, like the concepts of field developed by Michael Faraday (1791–1867), provides a context—a structure or pattern beyond the minute that requires reorganizing hierarchies but still urges the hunt for smaller elements. Only Einstein's early-twentieth-century rethinking of space and time suggests a different set of rules and presumptions, beginning with the ultimate transmutability of what had been elemental assumptions about the separate nature of space and time or matter and energy. Einstein's first insights about quanta were published in 1905 and, with the work of others such as Werner Heisenberg (1901–76), shifted physics away from the foundational notion of an elemental particle (though physicists continue to seek ever smaller particles) and formal concepts of complementary neatness—away from structuralism as a way of thinking.[10]

The Ultimate Significance of Strings

We cannot, however, forget the atom—believed in, predicted, charted for centuries, and still understood by the public of the twentieth century as the smallest basic element, itself composed of even (and increasingly) smaller elements. Joseph Proust (1764–1826) and John Dalton (1766–1844) recommenced work on the atom in earnest in the nineteenth century as a way to explain the regular proportions of elements found in chemical compounds. Proust later suggested that even atoms were comprised of smaller parts. In 1881 the British physicist George Stoney (1826–1911) named the electron; in 1897 J. J. Thomson (1856–1940) demonstrated that electrons are particles.[11]

By 1911 Earnest Rutherford (1871–1937) proposed a model of the atom like that of the solar system, with electrons orbiting around a nucleus made up of protons and neutrons. The Rutherford model still holds sway today in the popular imagination, though physicists have worked for a century to demonstrate the far greater complexity of the particle world. Accelerator physics (the branch of physics that studies the results of making particles collide at high speeds in an accelerator) has discovered more than three hundred, often rapidly decaying, particles. Physicists setting out to explain this variety of particles proposed yet other, more basic particles to account for those they had already discerned. Physicists propose twelve basic particles, divided into quarks and leptons.[12]

But even quarks and leptons are not the end of the minute basic particles. Attempting to bring together general relativity theory (the theory of gravity) and quantum mechanics (the rules governing the behavior of elementary particles), physicists such as John Schwarz (1941–), Michael Green (1946–), and Gerard t'Hroot have developed various string theories. Hypothesizing that everything is made up of minuscule vibrating fibers, string theory, according to Brian Greene, is "the ultimate explanation of the universe at its most microscopic level, theory that does not rely on any deeper explanation."[13] Differences in matter and force are caused by the different ways strings can be arranged (they can be open or closed loops) and vibrate. In addition, strings are understood to operate in ten spatial dimensions (only three of which we can see). The search for the answer to what the universe is and how it works continues to seek ever smaller units in the hope that eventually we will find a smallest common denominator, a particle that is common to everything and that enables a grand unified "M" or "Mother of All Theories" theory bringing together the various versions of string theory that currently exist.

That physics ends up with strings to explain how energy and matter relate to and transform one to the other and that it is currently looking for a "Mother" theory brings the physical sciences uncannily close to biology, which also has found its defining parental string or at least chain of acids. Although physics, cosmology, and biochemistry all arrive at the string as the representative of the conjoinder of matter and meaning, the string form also embodies an alternative way to think about order and meaning, a way not bound up necessarily in the binary structures of structuralism. The complexity of such infinitesimal systems no longer reiterates familiar structures such as the solar system but requires a leap of thought across multiple, mostly unseen, dimensions. The importance of small unifying bits, whether they are strings or DNA, gives way to the complexity of the environments and systems within which they operate. In the case of biochemistry, that complexity is simplified and obscured in popular accounts of DNA, where public acceptance is crucial and where DNA's imaginary operations provide narratives of legacy and cohesion through time as well as hope for intervention into the previously unresolvable problems of disease and mortality.

The irony of this epic search for the minute is that obviously smallness is not finally the only criterion. Rather, to be what we are seeking, the small needs also to both reflect and unravel the various orders and assumptions by which it was discerned. In 1869 Johann Miescher (1844–95) discovered what he eventually named "nuclean," a substance derived from the pus on bandages collected from Crimean war wounded. His teacher, Ernst Hoppe-Seyler (1825–95), did not think the substance important and refused to publish Miescher's article on it. Although in 1889 August Weissman (1834–1914) theorized that the chromosomes contained the material of heredity, the connection between Miescher's nuclean, genetic heredity, and DNA had to wait seventy-five years until 1944 when Oswald Avery (1877–1955) identified DNA as the "transforming principle"—as the chemical linked to heredity. That its structure, the mystery Rosalind Franklin (1920–58), Maurice Wilkins (1916–2004), Crick, and Watson solved, was a more momentous milestone than the discovery of the substance itself demonstrates the extent to which we regard structure as central. The coincidence of its structure with prevalent modes of thought—especially since knowledge of the existence of DNA long preceded its analysis—shows the extent to which science itself is not only bound to its context but bound to move toward the continuing coalescence of method and matter, of the materialization of the assumptions by which substances are adduced in the substances themselves.[14]

Although this would seem to suggest that science somehow reifies its own fantasies, what it may also suggest is that scientific methodologies and assumptions are simultaneously self-fulfilling and insightfully predictive. And just as all modes of knowledge reflect their context, so too do they move it beyond its own capacity to analogize. Cardinal objects such as atoms, DNA, and strings encapsulate both the modes of thought that exceed discoveries, resulting in pseudoscientific and magical claims, *and* the mechanisms—the new set of assumptions and dynamics—by which our understandings exceed the past. The object is, finally, a point of transition. Accounts of it are never really about it (DNA is not literally a "text") but about either its conceptual progenitors or its potential heirs—the imaginary of its extended context or the threats to context it may contain.

Form and Philosophy: Structure and the Nature of Being

If for millennia we have sought the smallest basic unit to explain what the universe is, we have not done so without also looking for ways to explain how basic units work together. The quest for the minute assumes that finding the "what" will provide both the where and the how. While structuralism has not always been the dominant mode of understanding the relations of matter and energy, structure has always been an important element in accounts of phenomena. When structure is understood to equal function, as it does in the ideas of late-nineteenth- and early-twentieth-century linguists and psychoanalysts, explanations become circular, phenomena are seen as self-sustaining (and almost self-explanatory), and the system becomes closed. This is not so much a danger to accounts of knowledge as it is to ways of thinking in general, for structuralism is a difficult mode of thought to escape.

In its specific applications, what we currently understand as structuralism grew from the scientific study of language inaugurated by the Swiss linguist Ferdinand de Saussure (1857–1913) at the turn of the twentieth century. Saussure's ideas depend on binaries—a dialectics that has a much longer history. Structuralism breaks language down to its smallest, most basic units, locating difference as the element by which these units are distinguished from one another and demonstrating how language operates through a system that repeats this basic opposition (or operation of difference) on every level.[15]

Saussure hypothesized that the relations between a word (or signifier) and what it means (or signified) are arbitrary. This opposition or dichotomy between word and meaning as well as between one word and another grounds a series of concepts about language's structure. First, language consists of both an abstract system (*langue*) and the actual practice of speaking the language (*parole*). Second, language use is divided between the selection of words or signifiers from among a number of similar possibilities and the combination of these words into sentences. These two processes suggest that language works along two axes: the vertical axis of word selection and the horizontal axis of sentence making. Following Saussure, the linguist Roman Jakobson (1896–1983) characterized these two axes as metaphor (word selection) and metonymy (word combination). In addition, Saussure posited that language works both diachronically, or through time, as its structure and signifiers slowly change, and synchronically, or in the way signifiers relate grammatically with one another at any given time. Thus Saussure offers three binary axes: one that runs between form or system (*langue*) and usage (*parole*), one that runs between word selection (metaphor) and word combination (metonymy), and one that runs between language change through history (diachrony) and variations in the present moment (synchrony). We tend to think of these axes, each in its own way, as vertical (or substitutions through time) and horizontal (or chaining at any given time).

Saussure's emphasis on binary oppositions was not a new way of thinking at the end of the nineteenth century but was itself an analytic method that echoed the dialecticism of nineteenth-century philosophy. The German philosopher Georg Hegel (1770–1831) developed an elaborate system relating the knowledge of "Being" with an inherent dialecticism—as the product revealed in the clash of two opposing ideas. We often conceive of arguments, for example, as proceeding from contrary positions to accommodation, as building from point to point to a conclusion that synthesizes conflicting views. But arguments also work the opposite way, breaking ideas down into their component parts. Thus even the familiar syllogism—All men are mortal; Socrates is a man; therefore Socrates is mortal—also works in reverse if the question is "What is mortality?" Dialecticism is a way to both arrive at solutions and find the underlying terms of the question.

Hegel's dialecticism focuses on how the revelation of Being requires breaking logic down into its component parts, or "aspects": an "abstract aspect," a "negative aspect," and a "positive aspect," the latter aspect providing the means by which Reason reveals the first two.[16] On a most literal level, this would mean something like this: we all know what the sun is as an abstract idea—as something the earth revolves around. We also know about the sun in relation to what the sun is not. It is not, for example, a planet or empty space. We know what it is not because we experience what it is—something that provides light and warmth. When we speak or think of one thing—the sun, say—we are depending on what that thing is not as a way to define what the thing is. Speech and thought reveal Being by bringing to light the dialectical nature of Being's underlying logic. At the same time, the underlying dialectic also reveals Being. Dialectic works both ways. It is a way to reveal "Being" as the object of scientific inquiry and is revealed in the expression of Being.

The dialectical is, thus, for Hegel the proper method for approaching scientific or philosophical thought. In Hegel's terms, scientific knowledge contemplates the object—the "Real"—without "preconceived ideas," without the kind of science (e.g., a Newtonian science) that tries to occupy a position outside the object or the Real and yet brings to the question many ideas about how the universe operates. This knowledge or philosophy ultimately, as Hegel's disciple and commentator Alexandre Kojève posits, "comes back toward itself and reveals itself to itself: its final goal is to describe itself in its nature, in its genesis, and in its development."[17] Though this refers to a desire for an unmediated knowing of the thing in its own truth, it also suggests that Being (or the object of study) reveals itself if we think about it properly. Of course, neither Hegel nor Kojève knew about genetic DNA, the substance that, it turns out, does this literally, its composition and structure defining its function and vice versa. For Hegel, this coming back—the idea that the object reveals the knower as it reveals itself—is itself an effect of the Real. Objects revealed also reveal that there are knowing subjects who can know them.

Hegel's understanding of Philosophy, or scientific knowing, suggests that Truth (Being or the Real) and Thought (or representation) exist in a self-generating, self-reproducing, self-sustaining dynamic. The object reveals a knower who reveals the object, ad infinitum. A knowing subject never stands outside the Truth or the Real. Rather, the dialectical relation among the Real, speaking, and knowing is generated from the Real as it is spoken—it is a condition of the Truth itself.

This dialectical dynamic does not, except in the most reductive fashion, anticipate and underwrite structuralism as it comes to privilege structure as meaning. Rather, it provides a model whereby dialecticism is already its own self-sustaining structure. In other words, by suggesting the interdependent nature of knowledge, truth, and Being, Hegel's philosophy anticipates how structure comes to be identified with function.

Hegel sets up a dynamically complementary model—a model where complementary functions and processes produce and sustain one another—that anticipates the kind of self-sustaining complementarity that characterizes Watson and Crick's epiphany about the nature of DNA.

Psychoanalytic Inquiry: Structures Make Sense

By the early to middle twentieth century, other disciplines such as psychoanalysis, anthropology, and literary criticism adopted a dialecticism derived from both Hegel's self-reflexive dialectic and Saussure's linguistic binarisms. Language became a structural analogy for larger organizations of sense making in culture, while ways of thinking about language and culture were already inflected by a rationalism and a scientific worldview. In *The Interpretation of Dreams,* Sigmund Freud, a contemporary of Saussure, produced a structuralist scheme uncannily like that of Saussure's theory of language. Freud saw dreams as working around two "governing factors": "condensation" and "displacement."[18] Condensation involves the combination or substitution of one figure, trait, or word for another; it recalls Jakobson's notion of metaphor, or substitution on a vertical axis. Displacement (or Jakobson's "metonymy") describes the dream's process of transferring traits from one element of a dream to another, which accounts for the dream's quality of distortion. Freud, thus, imagined that the human psyche itself was structured according to binaries similar to those proposed by Saussure for language.

Jacques Lacan (1901–81), a psychiatrist who followed and interpreted Freud's work, hypothesized that the "unconscious is structured like a language."[19] Lacan connected structure and psychic processes in a dynamic similar to the complementary interrelationship of Being and speech in Hegel: "The unconscious is constituted by the effects of speech on the subject, it is the dimension in which the subject is determined in the development of the effects of speech, consequently the unconscious is structured like a language."[20] Like the interrelationship of Being and speech in Hegel's thought, the unconscious, like Being, is produced by speech and in turn "speaks" the subject. Lacan's formula, like Hegel's, links Logos (speech) to a kind of "Truth" (the unconscious), which already embodies the structure of language. In Lacan's thinking, this reflexivity characterizes a kind of production/reproduction as a series of cybernetic transformations (cybernetic in that each change becomes the basis for any subsequent change), as the Symbolic (i.e., *langue*, the structure of language itself) produces a specific inflection in the unconscious that in turn reproduces the Symbolic itself, which produces the unconscious, and so forth.

In declaring that the "reality of the unconscious is sexual," Lacan also suggests how the unconscious shares its linguistic structure with our understanding of heterosexual reproduction, an essential connection that often implied a consistent, naturalized, circular rationale for dialecticism itself. The sexual reality of the unconscious looks like the binary structure of kinship systems and generational affiliation.[21] In other words, language's binaries are connected to, inform, and reflect the organization of culture into the gendered binaries of heterosexuality. Language and culture in turn rely on the imagined complementarity of sexual reproduction itself at even the most minute level or, in this case, the most unconscious level. This reproductive model reiterates the interchange between the general and the specific, the Symbolic and the individual. "Let us say," Lacan suggests, "that the species survives in the form of its individuals. Nevertheless, the survival of the horse as a species has a meaning—each horse is transitory and dies. So you see, the link between sex and death, sex and death of the individual, is fundamental."[22] Fundamental as well, then, is reproduction's link between species and individual. Imagined as the merging of two opposites into one, reproduction simultaneously produces more of the same through a chain of generations. Connected to structuring linguistic laws, reproduction becomes a basic premise of the unconscious as well as sustaining ideologies of capitalism and patriarchal organizations.

Repro-structure: Generation Mirrors Minds

Our ideas of reproduction, kinship, and language reinforce one another in a circular fashion. Reproduction, which according to Lacan springs between life and death, species and individual, is allied with "existence," which, "thanks to sexual division, rests upon copulation, accentuated in two poles that time-honoured tradition has tried to characterize as the male pole and the female pole."[23] If we believe that copulation constitutes "the mainspring of reproduction," the synthesis of two products into one, then we believe reproduction is the synthesis of binaries. Or, as Lacan suggests in his hesitation about definitive binary genders, via "tradition" we "try" to characterize male and female as binary, but this gendered dichotomy is in no way certain. Reproduction allied with the structure of language produces the need for a binary "time-honored tradition tries to fulfill." The convenient slotting of reproductive roles produces a more certain gender dialectic than actually exists. The binary gender that stars in each instance of reproductive activity also extends to and supports an entire social system, itself feeding the unconscious as a part of the Symbolic.

The idea of two parents, mother and father, taken in the context of all possible modes and practices of reproduction, is as much imagined as it is based on scientific fact. Many species reproduce without two parents, and many families persist through only partial kinship systems (i.e., no identified father, adoption, etc.) We maintain a notion of sexual reproduction in the face of what we call "asexual reproduction" in which organisms (usually "lower" species) have the ability to reproduce by splitting or fertilizing themselves. Current reproductive technologies such as cloning, egg-to-egg fertilization, or the use of multiple eggs promise the ability to bypass or alter heterosexual contribution. In other words, to assert the primacy of sexual reproduction, we must ignore other existing or conceivable reproductive practices. As Lacan suggests, this binary reproductive system, as natural as it may seem, is more an extension of the logic of the signifier—of language as binary—than a transcription of biological mechanism. We understand reproduction the way we do because of the way language is structured rather than because all reproduction necessarily involves the combination of sexual opposites. This omission of alternative reproductive modes is supported by the structure of language itself, which provides an overarching symbolic system that accommodates and privileges binary forms over all others.

The circle among language, the unconscious, and gendered binaries, however, isn't complete unless we also link, as does Lacan, this tendency toward binaries to the ways cultural practices express an overwhelming binary structure: "It is modern structuralism that has brought this out best, by showing that it is at the level of matrimonial alliance, as opposed to natural generation, to biological lineal descent—at the level therefore of the signifier—that the fundamental exchanges take place and it is there that we find once again that the most elementary structures of social functioning are inscribed in the terms of a combinatory."[24] The model of governing structures, in other words, is the sociocultural model of marriage, which follows the same rules as does language rather than model itself on any of the various biological mechanisms for reproduction. Combination, or the "combinatory," is, like language, structured as the joinder of complementary opposites. The fact that reproduction is imagined to require such opposites is an effect of the predominance of this binary structure.

Why, then, does this binary structure seem so ensconced—so natural? The idea of the "combinatory" links the structure of language to the biological structures of reproduction. These structures coalesce in "the signifier," which Lacan has defined in a discussion of the connection between the structural anthropology of Claude Lévi-Strauss (1908–) and psychoanalysis. "Before strictly human relations are established," Lacan asserts, "certain relations have already been determined. They are taken from whatever nature may offer as supports, supports that are arranged in themes of opposition. Nature provides—I must use the word—*signifiers,* and these signifiers organize human relations in a creative way, providing them with structures and shaping them" (emphasis mine).[25] The signifier, the provider of structure, is itself the product of the conjunction of the polar (binary) organization of society and the apparent binaries of biology such as male/female. Like Being and speech for Hegel, the signifier is the aggregate of symbol and body, neither of which is independent of the other, but mutually defining.

Lacan ultimately traces the operations of the signifier to the minuscule in the field of genetics:

> What would make it legitimate to maintain that it is through sexual reality that the signifier came into the world—that man learnt to think—is the recent field of discoveries that begins by a more accurate study of mitosis. There are then revealed the modes according to which the maturation of sexual cells operates, namely, the double processes of reduction. What is involved, in this reduction, is the loss of a certain number of visible elements, chromosomes. This, of course, brings us to genetics. And what emerges from this genetics if not the dominant function, in the determination of certain elements of the living organism, of a combinatory that operates at certain of its stages by the expulsion of remainders?[26]

The clarity and ineffable naturalness of the combination of binaries can only, however, be sustained by a constant exclusion or reduction of everything that cannot be crushed into oppositions or combinatory mechanisms. Even meiosis, for example, the process by which germ, or "sexual," cells persist with only one chromosome from each of the twenty-three chromosomal pairs, leaves half of an individual's genetic complement behind in another cell. This splitting results in what Lacan refers to in the passage above as a "reduction." To produce a new being, two germ cells combine, the chromosomes from two different organisms coming together. The sum of this is that there are two "reductions," one from each germ cell, and two "remainders" of genetic material left behind.

While DNA genes do not in fact work through the "expulsion of remainders" (as the production of germ cells might), the concept of the DNA gene as a reduction articulates precisely the position of the gene in symbolic systems. The gene is the imaginary embodiment of a binary principle never detached from ideas of gender, the logic of heterosexual reproduction, or the structure of kinship. It represents the crystallized intersection of the social and the biological, the symbolic and the material, species and individual, life and death, Being and speech, existence and signifier. The DNA gene is the perfect synthesis, Bacon's "intermediate," the signifier par excellence, whose significance reflects all other significance and whose imagined operation enacts the structuralist principle by which it is situated as a reduction of all.

From the Psyche to the Social

Lacan's understandings of how the signifier functions in the unconscious partially reflect the ways his contemporary Lévi-Strauss linked the workings of culture to linguistic structures.[27] Instead of seeing the rules of structure, or *langue,* as predominant, Lévi-Strauss reversed *langue* and *parole* (structure and usage), making *parole* (usage, the signified) the material of myth and understandings of the universe, and *langue* (structure, signifier) the social thought expressing those myths in one way or another. Lévi-Strauss's emphasis on the signified is important because it suggests a slightly different way to understand structure—not as determinative but as itself the effect of meaning, a process, or a structure located somewhere else. In his essay "Linguistics and Anthropology," about the intersection of structural linguistics and anthropology, Lévi-Strauss asserts that "we have been behaving as if there were only two partners—language on the one hand, culture on the other—and as if the problem should be set up in terms of the causal relations: 'Is it language which influences culture? Is it culture which influences language?' But we have not been sufficiently aware of the fact that *both* language and culture are the products of activities which are basically similar."[28] Lévi-Strauss locates these activities as features of the "human mind," which, if the similarities among cultural structures are not merely chance, must be in some part responsible for the similarity.

Unlike Lacan, who sees the "mind" as itself produced in and through linguistic, cultural processes, Lévi-Strauss allies the mind with a biology focused on binary gender as manifested in kinship structures as one site where language and culture, past and present intersect. In a move that merges structuralism and science, he reduces kinship to what he sees are its simplest elements—"the atom of kinship, if I may say so, when we have a group consisting of a husband, a woman, a representative of the group which has given the woman to the man . . . and one offspring."[29] To this four-element atomic structure (again curiously reminiscent of DNA), Lévi-Strauss applies another dichotomy, that of "positive behavior and negative behavior" as a way to understand the possible permutations of the kinship relations among the four possible positions he has listed.[30] This results in a series of rules about alliances and affinities based on these dichotomies. For example, when husband and wife have a positive relation, and brother and sister a negative one, then there is a positive relation between father and son and a negative relation between maternal uncle and nephew.[31]

What Lévi-Strauss's analysis demonstrates is the interinflectional complexity with which it is possible to understand structures when there are multiple layers, or "levels," of dialectic. In more "primitive" Hopi culture onto which Lévi-Strauss might suspect he has a less-biased view, several different binary structures coexist, each having a different relation to history depending on where in the larger structure a particular instance of structure is located. At one point in time, for example, the mother's line may be generational (it changes through history), the father's mother's line may be more or less atemporal (with its terms unchanging), and the "male ego" portion of the mother's line simply alternating between mother and father. This complexity, also a feature of Lacan's thinking, is perhaps the optimal result of bringing together structural linguistics and other systems of signification (or of even conceiving of kinship or the psychic as systems of signification).

Structuralist ideas about language and culture precondition conceptions of the gene as it is understood to constitute both heredity (or information through time, *langue,* diachrony) and individual form (or the expression of information, *parole,* synchrony). Up and until molecular biologists and others began trying to discern exactly how DNA transmits the information that produces proteins or knows when to do so, the structure of the gene (its *langue*) occupied the scientific imaginary. DNA structure, or *langue,* still arguably dominates the public imaginary where rules and regularity—structures—are more palatable (and salable) than the confusing complexities of actual genetic expression, or *parole,* or the even more incomprehensible (or simply nonavailable) extralinguistic analogies for DNA such as topography, meshed gears, or systems.

From Structure to Culture

The idea that the linguistic distinction between signifier and signified might be usefully deployed in thinking about other systems is explored by

Roland Barthes (1915–80) in a series of essays, *Mythologies,* in which he analyzes everything from Einstein's brain to professional wrestling.[32] In a later rumination on structuralist thinking, *Elements of Semiology,* Barthes maps out the rationale for what he identifies as the "semiological prospects" existing in discourses of code systems at "the frontiers of language."[33] Noting the affinities between Saussure's *langue/parole* distinction, the structural anthropology of Lévi-Strauss, and the sociology of Émile Durkheim, especially in the idea that cultures have a "collective consciousness dependent on individual manifestations," Barthes endorses the phenomenologist Maurice Merleau-Ponty's idea that "any process presupposes a system."[34] Enlarging the field to "any process" has meant, at least in the realm of sociological phenomena, the elaboration of "an opposition between *event* and *structure* which has become accepted and whose fruitfulness in history is well known."[35] This opposition becomes the basis of Barthes's own analysis of cultural phenomena: "There exists a general category *language/speech,* which embraces all the systems of signs."[36] In other words, even in the study of culture, the same binaries formulated by Saussure—form and individual instance (*langue* and *parole*), history and simultaneity (diachrony and synchrony), substitution and displacement (metaphor and metonymy)—operate.

With Barthes's readings, it becomes clearer that structuralism itself is not so much an extension of thinking about language as binary structure is itself an instance of a more pervasive structuralism—the kind of zeitgeist that prompted both Saussure and Freud to understand their subjects as effects of structure. Although Barthes continues to locate the source of structuralist thought in linguistics, he begins to see its manifestations as more general and more mimetic. In "The Structuralist Activity," Barthes suggests that "the goal of all structuralist activity, whether reflexive or poetic, is to reconstruct an 'object' in such a way as to manifest thereby the rules of functioning (the 'functions') of this object."[37] In other words, both criticism and art might compose themselves according to the insights of structure, producing what Barthes calls "a directed, *interested* simulacrum" that makes visible the structures not immediately discernible in the "natural object."[38] But Barthes takes this structuralist process a step further, proposing that structuralism is an "activity":

"Creation and reflection are not, here, an original 'impression' of the world, but a veritable fabrication of a world which resembles the primary one, not in order to copy it but to render it intelligible."[39] Structure, in other words, inevitably makes over the world in its own image, and the structure's image becomes the mode of access to the world.

Although structuralist activity for Barthes is primarily a critical and conscious endeavor through a range of disciplines and discourses (a range that very well could include the discourses surrounding science, though Barthes never takes his own work in that direction), his notion of "myth" offers another way to understand how a structuralist critical activity can identify and demythologize cultural ideologies. In the mid-1950s *Mythologies,* Barthes outlines the characteristics of "myth" as an organization of bourgeois ideology that "transforms history into nature."[40] While Barthes's work aims toward setting up a critical object and method based on the insights of structural linguistics, his insights can be turned around to suggest that the culture that is the object of structuralist study already, as Barthes intimates in "The Structuralist Activity," manifests the characteristics of dialectical structure. In other words, what structuralist critical activity discerns is the structure already operative in the systems it identifies and analyzes. The introduction or intervention of this structure, according to Barthes, occurs when "the world enters language as a dialectical relation between activities, between human actions."[41] The transformation of world into dialectical relations is catalyzed by "myth as a harmonious display of essences."[42] "A conjuring trick has taken place," Barthes suggests: "It has turned reality inside out, it has emptied it of history and has filled it with nature, it has removed from things their human meaning so as to make them signify a human insignificance."[43]

Even though structuralist critical activity can demythologize these myths (it can show their exaggerated binary structure and series of displacements), it may just as well be the case that structuralist activity itself is a part of that mythology, its process of discerning dialectical relations—the interplay of event and structure—itself naturalized, a conjuring trick whose most potent manifestation is the appearance of a critical discourse that sustains bourgeois ideology as it seems to reveal it. If the semiology of a code system forces the revelation of how its mythology naturalizes the historical relations that produced that system, then semiology naturalizes itself—situates itself as the obvious mode of decoding—in its operation. Not only is semiology, thus, a mode of analysis, it is also the potential object of its own process just as any other register of analysis or knowledge might become such an object. In the mid-1950s, for example, Barthes could have taken molecular biology and its stunning discovery, DNA, as the object of a semiological analysis, revealing in the process the mythology (the turning of history into nature) of science and in particular the gene. But the structural beauty of DNA is already, from the first declaration of its codelike quality, too much like the perfect functioning of a productive semiotics to be seen as myth—it is always already nature. This suggests that in Barthes's terms, DNA is also already bourgeois ideology.

All Paths Lead to the DNA Gene; or, The Performative End of Analysis

DNA and the gene are one site where structuralist activity and thought come together as the seeming truth and foundation of natural (biological) phenomena. They are also the site most broadly mythologized and most subject to bourgeois interests. It could be said, in fact, that the gene's mythologization enables its capitalization. But, too, the gene is already a part of the myth of an answer that fulfills the history of science—"the harmonious display of essences." In this sense, the concept of the gene as DNA becomes the self-referential nexus of a structuralism wrapped around itself. Structure as meaning ultimately leads to ideas of structure as function. It is what it does, it does what it is. This idea, called "the performative," also derives from insights about language. In a series of lectures in the mid-1950s, Austin identified a previously undefined category of utterances included in the general realm of the "statement" that he called "speech acts," or the "performative." In these "the issuing of an utterance is the performing of an action," such as when uttering "I do" constitutes a marriage.[44] In a narrow sense, the performative says what it does and does what it says in so saying; such statements have a legal or contractual effect. Austin delimits this characteristic to circumstances that fulfill certain conditions such as being in an appropriate context or being accompanied by other actions, such as hitting a ship with a bottle as it is named or placing a ring on the groom's finger and saying "I do."

This notion of the performative—of saying as doing what is said—has been taken more broadly and analogically beyond linguistic science, coming to refer to instances where language or any mode of signification (theater, film, plastic arts) produces the effects it describes or enacts in its saying. This performativity is often understood as one characteristic of postmodernism (a set of philosophies and aesthetic practices also named in the 1950s). But it is also the endpoint of structure, where structure meets itself, where structure is discourse and discourse is structure. When structure itself expresses—where structure is the same as event, where signifier and signified appear to be the same thing—there appears to be no gap or arbitrary relation between signifier and signified: the arbitrary connection, for example, between the word *tree* and the object it designates. Arbitrariness (which is still present in the linguistic signifier as well as in the ritual nature of performative utterances) bows to function, which, in producing the illusion of secure reference ("I do" = marriage), also produces the self-identity of signifier/signified. The appearance of secure reference—this signifier always refers to this and only this signified—is produced as an effect of retroactivity. Retroactive signification means that we apply what we know at the end of a statement to elements at the beginning. The relation produced by the performative (the marriage, contract, wager) is secured by the relation created, even if, in order to produce such a relation, the reference—what we think language means (always a tricky business)—must be certain in the first place.

The apparent closed-circuitry of the performative provides the illusion of a perfectly symmetrical and self-enclosed system where language is the doing and the doing is the language. The category of the performative provides both model and instance of the crystalline self-identity of structure and expression, which in embodying the most perfect structuralist moment also short-circuits its analysis. If a statement does what it says, how does one know the difference between its doing and saying? Which is language and which is act? What is structure and what is function? How does one demythologize the myth of structuralist perfection if the strands available for analytic grasp are completely interwound? How does one locate a dialectical structure in a sphere?

The philosopher and critic Jacques Derrida undertook a response to the claims of Austin and his followers in the 1970s' interrogation of the possibilities of reference and communication in the collection *Limited, INC.*[45] Showing the assumptions about representation that undergird Austin's notions of the efficacy of speech acts, Derrida demonstrates how signification depends on constant deferral, iterability, and absence. As Austin formulates the performative as an instance in which linguistic signifiers coalesce with their signified—an idea that ultimately asserts both nonarbitrary reference and the coincidence of meaning and materiality in language—Derrida shows that such coincidence is itself an instance of deferral and iterability. In other words, even performative speech signifies the absence of something—connection, deferred meaning—and requires iterability, the ability to restate without repeating, since every utterance occurs in a different context.

Derrida's intervention demonstrates the phantasmic hope represented by the category of the performative, a hope for language's material effect and coincidence with meaning. The figurative language with a similar material effect, DNA—or its textual and linguistic analogies—invites a similar critique. There is no self-identicality. Nothing is ever repeated, only iterated, each time in a different context, looking away from structure to more systemic understandings of iterated instances.

Erwin Schrödinger (1887–1961) made the last great attempt to reconcile physics and biology as a single set of processes based on structure—on how the laws of particle physics might govern the behavior of molecular biology. In a series of lectures he delivered in Dublin during World War II, Schrödinger ruminates on the permanence of the organic "code-scripts" by which life is preserved and perpetuated. Working from the insights of Heisenberg and presuming a continued discovery of ever-smaller particles, Schrödinger tries to think through how "events *in space and time* which take place within the spatial boundary of a living organism" can "be accounted for by physics and chemistry."[46] He postulates that "a well-ordered association of atoms, endowed with sufficient resistivity to keep its order permanently, appears to be the only conceivable material structure that offers a variety of possible ('isomeric') arrangements, sufficiently large to embody a complicated system of 'determinations' within a small spatial boundary."[47] In other words, Schrödinger predicts the existence and molecular qualities of DNA. Schrödinger's lectures-turned-book *What Is Life?* hit the right audience. Watson credits it with changing the direction of his research: "I spotted this slim book in the Biology Library, and upon reading it was never the same."[48]

Schrödinger connects "the mechanism of heredity" to "quantum theory": "In the light of present knowledge, the mechanism of heredity is closely related to, nay, founded on, the very basis of quantum theory."[49] Hypothesizing that "biological stability" is connected to "chemical stability," Schrödinger suggests that the quantum theory of Max Planck (1858–1947)—the fact that energy is transmitted only in certain-sized packets, or "quanta"—might explain the stability of biochemistry. Observing that Planck formulated quantum theory at the same time that Mendelian genetics was rediscovered (1900), Schrödinger thinks that any connection between the two could "emerge" only when both genetics and quantum physics had reached a "certain maturity."[50] The idea that small particles occupy specific "states," or energy levels, explains why it is that "small-scale" systems change discontinuously, or through quantum leaps. Although, as Schrödinger explains, quanta states are far more than merely energy levels, molecules such as are present in living organisms exist through generations in a state of constant temperature; there is nothing that could excite these molecules to change their "state," hence their constancy.

But despite his enlistment of quantum theory, Schrödinger is most fascinated by the permanence of the code-script function he ascribes to the chromosomes. His thinking about this code-script, despite his quantum explanations, focuses on it as a structure in a way very much like the ways structuralists think about structure, particularly performative structures: "The chromosome structures are at the same time instrumental in bringing about the development they foreshadow. They are law-code and executive power—or, to use another simile, they are architect's plan and builder's craft—in one."[51] We might trace the contemporary insistence of the code metaphor to Schrödinger's pre-DNA formulation. But we might also understand Schrödinger's characterizations as being themselves affected by the ways structuralist thought dominated understandings of expression even before the insights of the 1950s.[52] Schrödinger's insistence on the expressive quality of the chromosomes is a comparison that in itself evokes the linguistic metaphor he comes to employ.

Schrödinger, like earlier atomists, reduces life to the gene, to the permanence of a code-script molecule that resists change and consistently directs the production of life through time. Watson saw Schrödinger's equation of genes and life—"Schrödinger said the essence of life was the gene"—as the "luck" that set him on the path that led to the discovery of the structure of DNA.[53] Schrödinger's hypothesis of a connection between structure and function, between permanence and expressiveness, between "plan" and "craft," cannily maps the way the larger world will understand DNA, genes, and heredity as first and foremost a text, language, or representation of some kind. These metaphors, which work structurally rather than systemically, provide a structuralist conception of DNA and genes at the very moment when they might have been portrayed as parts of complex, organized systems.

Michel Foucault addresses the questions of why structuralist systems are resistant to analysis and the reasons they are pervasive, suggesting that structuralism as an analytic method can only ever reproduce its own terms. Like the contemporaneous Derrida, Foucault questions the assumptions subtending the meaningful self-reproduction of structuralism. If structuralism is the method, then structuralism is the product. Even asking what might be outside or different from structure is a question produced by and through structuralism. In this sense, the gene as the perfectly fulgurating structure not only represents an admirable resistance to its own demythologization but becomes the apotheosis of a fantasy of structural beauty in a larger sense, engendering and sustaining structuralist thought at the point where it also begins to crumble: the mid-1950s. In the discourses of knowledge—in the modes available for analysis—the concept of the gene secures the functionality of structure as a way of thinking. DNA makes it difficult to conceive of structures as anything other than perfectly and ideally complementary. In this sense, DNA conserves the systems through which it emerges.

And it is not as if those thinking about available modes of thought were not aware of structuralism's effects. In *The Archaeology of Knowledge,* for example, Foucault demonstrates how the myth of structuralism produces a conundrum something like the performative—or DNA. In his attempt to set forth an "archaeology" of the "discursive formations" of knowledge (as opposed to a "history of ideas"), outlining the conditions "with which a practice is exercised," Foucault simultaneously battles a persistent structuralism.[54] The structuralist avatar makes it difficult to conceive of any category in a way that does not reflect structuralist precepts; even conceiving of a nonstructuralist system requires thinking about structuralism.

In *The Archaeology of Knowledge* Foucault's project is to formulate a nonstructuralist way to think about knowledge, especially to avoid the inevitable structuralism introduced by the categories of language (on which structuralist ideas were based in the first place). How can we even conceive of a field if our conception of the object of study itself already depends on the presuppositions by which it is selected? A good example is the analytic methodologies of the scientific method wherein a search for basic components leads to the discovery of those components (such as atoms or DNA) as the answer or truth. Seeing the molecule as the answer to heredity not only answers structure with structure, it also substitutes a substance for a process.

To get at modes of analysis as such, Foucault must differentiate his concepts from those of structuralism. "Far from being the principle of individualization of groups of 'signifiers' (the meaningful 'atom,' the minimum on the basis of which there is meaning), the statement is that which situates these meaningful units in a space in which they breed and multiply."[55] Rather than base the identification of "discursive formations" on the basic structuralist opposition between signifier and signified—the "atom" of structuralism—Foucault wants instead to locate the relations through which "meaningful units" can proliferate, relocating the issue from structure to production, from text to context. Trying to avoid precisely the problem of the "theoretical choice," or mode of analysis, that predefines possible answers or terms of analysis, Foucault sets out "to show how a domain can be organized, without flaw, without contradiction, without internal arbitrariness, in which statements, their principle of grouping, the great historical unities that they may form, and the methods that make it possible to describe them are all brought into question."[56] Trying, in other words, to merge method and object, Foucault adopts a more systemic approach to avoid the oppositional structuralizing that reiterates itself. "I am not proceeding," Foucault declares, "by linear deduction, but rather by concentric circles, moving sometimes towards the outer and sometimes toward the inner ones."[57]

Among the domains Foucault identifies as discursive formations is the domain of science, which, he admits, works a little differently than the "archaeological territories" he has been mapping, in that science maintains "certain laws of construction" governing its propositions.[58] In other words, because science, like law, is interested in the truth-value of its propositions, it appears to be invested in a certain consistency of method—the scientific "method." But such investment does not mean, as Foucault proposes, that scientific discourse is free from ideology. Rather than work at the level of the structures discovered, their utility, or even in the consciousness of scientists, ideology appears in science

> where science is articulated upon knowledge. If the question of ideology may be asked of science, it is in so far as science, without being identified with knowledge, but without effacing or excluding it, is localized in it, structures certain of its concepts, systematizes certain of its enunciations, formalizes certain of its concepts and strategies; it is in so far as this development *articulates knowledge, modifies it, and redistributes it on the one hand, and confirms it and gives it validity on the other* . . . it is the question of its existence as a discursive practice and of its functioning among other practices. (emphasis mine)[59]

In Foucault's terms DNA represents such a nexus or point of articulation of science and knowledge, almost more literally than Foucault intends in his own formulation. It certainly articulates knowledge in the sense that we believe that DNA contains the complex set of instructions for life processes. DNA modifies knowledge as well, both as a part of its process and as an effect of its own reduplication. It certainly redistributes knowledge from generation to generation and cell to cell. At the same time, the discovery and presence of DNA confirms both identity and the validity of the science that discovered and wishes to wield it.

The ideology here, as Foucault suggests, does not exist in the truth-value of DNA as an agent of heredity or a molecule that reproduces itself but in the ways DNA—the point of articulation between science and knowledge—is understood to function like a discourse, in fact, like a language. The linguistic structures through which DNA is characterized and understood, linked absolutely to the chemical functions DNA represents and as such taken as fact rather than analogy, also reaffirm the structuralist ideology of science in the 1950s, an ideology that, for various reasons, is already evolving away from structure in the 1950s but persists even now.

The Loop

Foucault's excursions into nonstructuralist modes of thinking are important in that they reflect part of the context within which DNA emerges. Science, as Foucault describes it, is a discursive practice that functions "among other practices," among which are emerging nonstructuralist modes of analysis that could just as easily be used to define and describe DNA. The question is not why DNA is taken as structure par excellence but why it isn't understood in more systemic terms from the start, especially since it is clear that DNA never acts by itself or only synchronically but in concert with a complex set of processes and relations through time. The fix on structure and the beautiful complementarity of its elements belie the necessary complexity of what must be multiple functions from conception through fetal development to aging.

Almost from its beginning, the twentieth century devised other ways to think about phenomena. Einstein, for example, began his hypotheses in 1906, hypotheses that no longer depended on an underlying set of binary oppositions. Systems theory and cybernetics, for example, were already available for those working with DNA in the early 1950s.[60] Both systems theory and cybernetics emerged contemporaneously with the structuralist applications of the mid-twentieth century in anthropology, psychoanalysis, cultural theory, and linguistics (Lévi-Strauss, Lacan,

Barthes, Austin) as well as with the development of digital computers. In tandem with the increasing complexities of quantum physics, systems theory and cybernetics address ways of understanding complicated, multilayered, nonlinear dynamics, the flow of information, and the intrinsic part played by communication and observation. Echoing the ideas of Heisenberg, whose Uncertainty Principle declared both that one cannot know the speed and the position of a particle at the same time and that the observation of phenomena was an integral part of what was being observed, systems theory and cybernetics included the element of the observer and the flow of information in models that accounted for the interrelation of structures that changed over time.[61] Observed phenomena included the variable of the observer, and therefore no "objective" view was possible. The laws governing individual instances, if such instances were unobserved, produced only sets of possibilities, "fields" of probability rather than any definable or predictable certainty.

All of this came together in the development of systems theory. Systems theories, according to Gregory Bateson, represent a "new way of thinking about what a *mind* is."[62] The four "essential minimal characteristics of a system," according to Bateson, are

> (1) The system shall operate with and upon *differences*.
> (2) The system shall consist of closed loops or networks of pathways along which differences and transforms of differences shall be transmitted (what is transmitted on a neuron is not an impulse, it is news of a difference).
> (3) Many events within the system shall be energized by the respondent part rather than by impact from the triggering part.
> (4) The system shall show self-correctiveness in the direction of homeostasis and/or in the direction of runaway. Self-correctiveness implies trial and error.[63]

What Bateson describes is a system in which change or movement can come from many different sites instead of one primary designated site (the "triggering part") and in which all parts can correct their behavior in relation to too much sameness ("homeostasis") or too much activity ("runaway"). The system, thus, is a way to organize differences into networks whose relation to one another is nonlinear (there is no set starting point) and which respond to feedback (what the system has already done). Instead of understanding phenomena as analytic or mechanistic (structuralist) approaches do—as set structures in which the relation of smaller component parts account for behaviors—systems theory understands the interrelation of parts and actions as elements of multilayered, looping, dynamic networks.

The systems theory described by Bateson is closely related to concepts of cybernetics also developed at the midcentury height of structuralist endeavors. Cybernetics focuses on the feedback aspect of systems. Modeled partly on the operations of the human nervous system, cybernetics is a mode of self-correcting regulation with pragmatic applications in control systems, computers, prostheses, and other mechanisms requiring sophisticated modes of self-correction. As its developer Norbert Wiener explains, "When we desire a motion to follow a given pattern the difference between this pattern and the actually performed motion is used as a new input to cause the part regulated to move in such a way as to bring its motion closer to that given by the pattern."[64] In this sense cybernetics proposes a circularity, or recursivity, in which what happens the first time (e.g., how close a hand reaching for an object comes to that object) becomes part of the information that governs how the hand might reach for the object a second time. The result of the first action becomes the basis for the second.

Systems theorists like Ludwig von Bertalanffy understand the development of systems theory as "a change in basic categories of thought of which the complexities of modern technology are only one—and possibly not the most important—manifestation. In one way or another, we are forced to deal with complexities, with 'wholes' or 'systems,' in all fields of knowledge. This implies a basic re-orientation in scientific thinking."[65] Tracing the origins of systems theory to the 1920s, von Bertalanffy understands systems theory as the need to manage the complexities of technology and multiple intersecting organizations. It also, as he points out, is an attempt to think profitably about "problems of order, organization, wholeness, teleology, etc." that "appeared central" to such fields as biophysics but that "were programmatically excluded in mechanistic science" (science that, as I have shown, focuses on the structural relation of parts).[66]

J. D. Bernal goes a bit further, seeing systems theory as the whole new way of thinking needed to bring together the rapidly expanding insights of quantum physics and astrophysics:

> Something radical is needed, and it will have to go far wider than physics. A new world outlook is being forged, but much experience and argument will be needed before it can take definitive form. It must be coherent, it must include and illuminate the new knowledge of fundamental particles and their complex fields, it must resolve the paradoxes of wave and particle, it must make the world inside the atom and the wide spaces of the universe equally intelligible. It must have a different dimension from all previous world views, and include in itself an explanation of development and the origin of new things.[67]

Although Anthony Wilden mounted an extended critique of systems theory as itself structuralist in the early 1970s, work in nonbinary, nonhierarchical ways of thinking took hold in more poststructuralist modes of thought such as Niklas Luhmann's work in systems and Gilles Deleuze and Félix Guattari's ventures into complexity, as in *A Thousand Plateaus.*[68] In considering the impact of systems theories, Cary Wolfe concludes that what is important about systems theory "is its ability to mobilize the same theoretical apparatus across domains and phenomena traditionally thought to be pragmatically discrete and ontologically dissimilar, *while at the same time* offering . . . a coherent and compelling account of the ultimate contingency of any interpretation or description."[69]

Systems theory in general provides an alternative paradigm for understanding the complex dynamics by which various phenomena interrelate. It emerges roughly contemporaneously with the description of DNA and offers a different way to understand how parts and wholes and multiple systems work together. No longer fixed on the smallest part as the answer to questions about the essence of life and the universe, systems theory shifts focus from structure to organization and interrelationship, seeing no single point as necessarily elemental or primary and including observers and operators in their understandings of how systems work. Systems theory incorporates and organizes elements from seemingly disparate fields—mechanics, biology, quantum physics, mathematics—reorganizing our way of conceiving fields and phenomena not as separate hierarchies, each working on the basis of a primary unit (the cell, the atom), but as fields that connect and affect one another. On the one hand, this shift represents an urge toward unifying theories—accounting for all phenomena within one kind of system. On the other hand, it suggests that we have outgrown traditional notions of disciplinary fields, which can no longer be regarded as independent from one another.

The urge to pull together disparate theories—particle and quantum theories, relativity, biology—represents part of the process Thomas Kuhn calls a "scientific revolution." Scientific revolutions are, according to Kuhn, "non-cumulative developmental episodes in which an older paradigm [in this case reductionism and structuralism] is replaced in whole or in part by a non-compatible new one."[70] The mid-twentieth century was faced not only with technological advances but with the rising desire to find a way to understand how the behavior of subatomic particles might relate to molecular biology and how all of that might fit with Einstein's theories of relativity—in other words, how different versions of the very small relate to one another, to the apparently Newtonian universe we inhabit, and the cosmology we only begin to understand at one and the same time, while also accounting for the observer. The need to discern theories that unify what were perceived as disparate regimes such as, for example, biology and physics produced various theories of system that either sought some common underlying element (in the way Schrödinger looked for atomic answers to heredity) or moved from field-specific analyses to attempts to systematize larger notions of "order, organization, wholeness, teleology, etc." that we see in systems theory.[71]

Protocols

This very condensed rough history of ideas tells us several things:

First, genes and our conception of their structure mark and conserve the coincidence of three strains of thought: an ancient strain of atomism or reductionism, the scientific method of the Enlightenment, and the twentieth century's focus on structure.

Second, although there may have been many ways to arrive at DNA and many ways to characterize it, the metaphors that characterize DNA—such as a book, an alphabet, a blueprint, or software—are especially resonant within the structuralist framework that produced them. Although scientists engaged in DNA research long ago elaborated on and departed from the structuralist models that dominated mid-twentieth-century thinking, representations of DNA operation aimed at the general public today have retained their structuralist and primarily textual and linguistic quality. We might understand this continued appeal to DNA as language or text as a simple simplification—as an attempt to convey a complex process in familiar terms.

Third, retaining these analogies, however, has several less felicitous effects. The repetition of metaphors such as "the gene is like a book" urges the public to keep thinking in the outdated and comforting terms such comparisons evoke, which in turn allays suspicion and fear, encourages acceptance, fosters hope, and discourages more probing questions and the necessity of inquiry. DNA's metaphors are not a coincidence; they conserve a particular way of seeing and understanding the world in the face of more complex and unfamiliar possibilities. For example, when we think of DNA and genes as codes that work like language, we also conceive of them as binary, meaningful in their arrangement, and manipulable, resonating ideologies of kinship, reproduction, and the natural hegemony of the complementary and the heterosexual. If DNA is a language, we can make it say things. We can control it, not it us. These qualities provide a useful platform for many other kinds of social, economic, and ideological uses. For example, the cultural binaries of gender, race, and sexuality embedded in our notions of language and text are perpetuated as unquestioned parts of structuralist interpretations and analogies. This in turn conserves fairly traditional notions of these categories as "true" or scientifically based rather than as themselves already products of a particular way of thinking belied by the discoveries of biology. If, for example, assumptions about gender influence the traits we look for as associated with gender, it is likely that we will find them, not because they are there but because we are posing the question in ways that permit only certain answers. Looking for a "gay" gene already presupposes that homosexuality is somehow a constitutional disposition hardwired into a genetic code. This presumes that gayness occurs in a single identifiable form and is the same for both males and females. Again, what constitutes the homosexuality one looks for reflects ideologies about sexuality current in the culture.

Fourth, through these analogies, genes take on the qualities necessary for capitalization and ownership, as textual metaphors urge us to think of DNA and genes as authored products. The analogies mask the commodification of processes and elements not typically thought to be commodities. It is not so much that now life can be bought and sold (this is the logical extension of the metaphor "DNA is the book of life") but that the various processes and subproducts of DNA research can be thought of as separable and patentable products and processes.

Fifth, attributions made to the power of genes are quite similar to the kinds of triumphs credited to miracle cures. Genes are treated as shortcuts much like the magic tonics imagined in mid-twentieth-century science films. Finding a gene for a disease is presented as if science has found an on-off button or a talisman. If, for example, wearing a copper bracelet works immediately and effectively on arthritis pain, so locating the gene for Alzheimer's disease is heralded as the next step to a complete cure. Paradoxically, announcing the discovery of "disease" genes has the effect of perpetuating pseudoscientific ideas as another of the many conflicting side effects of publicity processes. More important, instead of inspiring citizens to experimental science, it confirms their belief in a more powerful and instant science that gives humans shamanistic control over complex phenomena.

Sixth, the ways concepts of genes reflect past intellectual history rather than ways of thinking contemporaneous with their discovery (structure instead of system) suggest paradoxically that the discovery and our subsequent conceptions of DNA are predisposed toward the past rather than the future. What this means is that in attempting to project the effects of the discovery of DNA and the listing of the genome, we tend to think in terms of what we can already do and the methods by which we accomplish it rather than see the ways complex systems might change our way of thinking about problems altogether. This is parallel to the anachronistic tendencies of science fiction. Instead of thinking, for example, about engineered DNA as altering a portion of a complex system, we think of it as solving single problems, such as improving the vitamin intake of poor populations by adding a gene for beta-carotene to rice. Such piecemeal alterations reflect the ways we think of science within a narrative rather than a systemic paradigm, or, at least on the surface, the description and rationale for the adoption of genetically modified agricultural products are represented within a narrative (simple cause-and-effect) paradigm. The problem—poor nutrition—is the evil part of the story to which we must respond. The solution—altering DNA of food staples—is produced only in relation to the problem, using science (in whatever state of advance it might be) to alter or respond to the problem, producing a solution, victory, salvation, as the poor of the world are provided with vision-saving vitamins.

It is, of course, no wonder that there are objections from anyone who understands that the world is not so simple either biologically or economically. Some scientists become frustrated at the public's lack of acceptance of new technologies. "Let me be utterly plain," protests Watson, "in stating my belief that it is nothing less than an absurdity to deprive ourselves of the benefits of GM [genetically modified] foods by demonizing them; and with the need for them so great in the developing world, it is nothing less than a crime to be governed by the irrational suppositions of Prince Charles and others."[72] Demonizing or criminalizing differences of opinion is less a question of right and wrong than it is an effect of corporate science's simplistic public descriptions in the first place. Public reluctance is both an effect of the obvious arrogance of science and the memory of other such "harmless" tinkerings with the biosphere (like DDT or the massive use of antibiotics in livestock) that indeed have had deleterious systemic effects. If we believe DNA is so basic and powerful, why wouldn't we object to its piecemeal application to specific problems? The clash over applications of genetic technologies in any other than a medical realm is a clash of narratives, where, perhaps paradoxically, the "nonscientists" take a more systemic approach to science's narrative sales pitches.

Seventh, as the means by which we have come to distinguish individuals, however, contemporary uses of DNA identification also emphasize the minute differences among individuals, instead of, for example, reiterating the surprising commonality of genes among individuals and even species. In its forensic use, DNA, though inconceivably complex and myriad, is reduced to a few representative sites or markers of difference that come to stand, at least evidentially, for all of the human genome. This spreads the impression that somehow we have scanned the entire genome for the special combination that makes this individual or that completely different from everyone else. In turn, ideas of difference become increasingly keyed to the proteomic differences produced by DNA. So blood types are a significant difference, race is not. Gender is a significant difference, age is not.

Eighth, if we combine the idea that genes have survived through evolutionary millennia (we share genes with yeasts) with the idea that genes determine individual identity, we get the idea that somehow individuals are potentially immortal. This chimera of individual survival is already a part of the fantasy of reproduction as a means of extending self by extending the genetic line. Just as genes are envisioned as a panacea for disease that might extend lives, the genome becomes the extension of lives through generations. If finding a gene for a disease implies the promise of a cure, so the survival of an individual's genes implies individual longevity through the survival of DNA itself. DNA constitutes both individuality and immortality, and DNA makes it but a tiny leap from the one to the other.

Finally, in the ways DNA represents a history of ideas and underwrites a variety of cultural values and beliefs, DNA functions as a mirror of the ways Western culture thinks about knowledge and itself. This DNA mirror is notably flexible and chatoyant, able to reflect multiple and conflicting ideas, attitudes, and controversies. DNA's beauteous structure functions as a medium flexible enough to absorb the various changing ideas about how meaning is produced and communicated, what and who is important and central in Western culture, and how old ideas are conserved through their artful transmutation into the new. In other words, DNA may be the structure through which genetic material is reproduced and disseminated, but it is also a mechanism for reproducing and disseminating thinking (ideologies) in the name of science and truth. And those ideas are reflected in the very terms employed to describe DNA for nonexpert audiences. DNA shows us the mechanisms by which Western culture negotiates new ways of thinking and technologies, as it is the pivotal concept for introducing the new as the old. And yet, in its constant iteration, DNA is never merely a reflection or repetition, but the threatening seed of something else whose potential emergence is persistently damped by analogies that appear always to link DNA to meaning, matter to life, Enlightenment science to the secrets of being.

CHAPTER 3

Flesh Made Word

I don't know if people realize that we just found the world's greatest history book. We are going to be up every night reading tales from the genome. It's so cool.
—Eric Lander

Genes are just chunks of software that can run on any system: they use the same code and do the same jobs. Even after 530 million years of separation, our computer can recognize a fly's software and vice versa. Indeed the computer analogy is a good one.
—Matt Ridley, *Genome*

High Style

As James Watson recounts it, the moment of insight was almost apocalyptic: "Upon his arrival Francis did not get more than halfway through the door before I let loose that the answer to everything was in our hands."[1] By lunch that day, Watson reports, "there was also the too obvious fact that the implications of its [DNA's] existence were far too important to risk crying wolf. Thus I felt slightly queasy when at lunch Francis winged into the Eagle to tell everyone within hearing distance that we had found the secret of life."[2]

This story provides a clue about how we have regarded DNA ever since. It is "the answer to everything," "the secret of life." At the same time we are queasy about something, worried perhaps about hubris, not quite sure that we want to make the final pronouncement. Just as DNA's matching pairs make it able to reproduce itself, our representations of DNA somehow reproduce ourselves—our anxieties, contexts, ways of thinking. As descriptions of DNA spread through culture, this self-reproduction occurs on the level of both molecular biology and cultural ideology. If the DNA gene is both the perfect and predictable product of a long history of ways of thinking about phenomena and structure, it is also an elastic reflection of how we currently perceive our own powers and abilities and fantasize a future in a postdigital culture.

One manifestation of DNA's cultural reflectivity (or self-reflectivity) exists in the ways we come quickly to perceive DNA as the answer without understanding either what the question actually is (is it What is a gene? or How do genes transmit heritable traits? or Why does DNA only occur in cell nuclei? or What is the role of DNA in cells?) or how DNA manages to accomplish much beyond reproducing itself. For example, Watson and Crick's first paper on the structure of DNA, published April 25, 1953, in *Nature,* modestly announces that DNA structure "has novel features which are of considerable biological interest."[3] One feature, they subtly divulge at the end of this initial article, is that "the specific pairing we have postulated [thymine with adenine and cytosine with guanine] immediately suggests a possible copying mechanism for the genetic material."[4] In their second *Nature* paper one month later in May, they proclaim that "the importance of deoxyribonucleic acid (DNA) within living cells is undisputed."[5] The one month from "considerable interest" to "undisputed" importance is prophetically typical of the kinds of instant knowledge and imminent cure attributed to DNA genes ever since, despite the fact that scientists themselves are careful to warn that apart from the self-replicating mechanism of DNA, we don't really know how genes get to be traits and propensities and behaviors.

Another manifestation of DNA's cultural reflectivity is how this instant notoriety is produced by and produces a particularly spectacular point of self-contained miracle. DNA is what it is because it is what it is, and it is what it is somehow instantly, completely, overwhelmingly. In an essay in *Nature* authored by Francis Crick to honor the twenty-first anniversary of the April 1953 paper, he muses:

> There is a more general argument, however, recently proposed by Gunther Stent and supported by such a sophisticated thinker as [Sir Peter] Medawar. This is that if Watson and I had not discovered the structure, instead of being revealed with a flourish it would have trickled out and that its impact would have been far less. For this sort of reason Stent had argued that a scientific

discovery is more akin to a work of art than is generally admitted. Style, he argues, is as important as content.[6]

"Style" in the discovery of DNA is about a flourish, about a Joycean turning point epiphany in which the truth is not there one moment, all there the next.[7] But what is really artistic about the epiphanic quality of DNA's discovery is how this insight, gained from painstaking model building that cleverly patched together the work of some rather more forgotten researchers (such as the sadly maligned Rosalind Franklin and Maurice Wilkins, cowinner of the Nobel Prize with Watson and Crick), actually enacts a shift from the styles, presumptions, and ideals of the modernist world that preceded it to the performative pastiche of postmodernity that quickly follows.

DNA is style where style is about the old in the new, the past in the future. DNA's precise job is, in fact, to provide the means to convey the past into the future. Heredity in self-reproduction, symmetry in the complementarity of its constituent nucleotides, a pastiche of ancient strings, junk codes, and functioning segments, DNA genes embody, recirculate, and reflect not only human reproducibility but also the style of their own ascension as an idea, the cultural moment of their discovery, and the prolonged oscillations between modernism and postmodernism, structure and performativity, art and science, certainties that DNA is central and uncertainties about how it works that have characterized the fifty years since DNA's discovery. Stent's observation was laudably canny. DNA has become the style: a style in which style is important, where the medium is the message, where essence is contained in a resounding chain of self-replications from the smallest cellular scale to the sweep of history.

In the Beginning Is the Word

The familiar figures of code, blueprint, book, map, alphabet, digital language, recipe, instructions, software, autobiography, history book, and even a Chicago gangster populate the genetic landscape with visions of libraries of manuals overseen by a scrivening molecular homunculus. As what we now perceive to be the most minute site of life's operations,

DNA is the animating, originary Word made flesh. The primacy of textual analogies of various sorts—from codes to letters to history books—suggests that DNA is imagined as something that stands for something else in the same way that words or images refer to concepts or objects by representing them. Representation means basically that we have agreed culturally that certain symbols stand for and refer to certain other sounds, objects, ideas, feelings, and relations. Hence the letter *t* stands for a certain sound, the symbol + means that we are adding something, the word *love* or the image ♥ refers to feelings of fondness, and the word *car* refers to a vehicle used to transport people and things. Apart from the image of the heart (which is itself only a rough approximation of an organ believed to be the seat of feelings), there is no intrinsic or necessary relation between the symbol used to represent something and the something such a symbol represents. The relation is arbitrary.

Almost everything is the product of representation in that we deal with the world through language. Even the metaphors of the map or blueprint are representations in that they render three-dimensional objects on a two-dimensional plane through a series of conventions by which we recognize what they depict. As in language, these conventions—scale; rendition of the earth's curve; the ways roads, rivers, and railroads, or plumbing, walls, and landscape, appear—are arbitrary. Their imagistic quality makes us think of them as more representational and less arbitrary than linguistic texts, but without conventions by which elements are represented, maps and blueprints would make no sense.

What's in a Word?

The analogies employed to describe DNA genes evolved from Charles Darwin's idea of the "factor" to Gregor Mendel's notion of the "element" to genes and DNA as templates and "codes" to comparisons between DNA and language, text, or software, especially after software became a familiar public concept (though members of the public who actually try to use certain brands of software might worry a little about any real parallel). The genome itself has been most grandly extolled as a book, a story, and a history. Occasionally, both DNA and genes have been conceived as little personified agents or homunculi with wills and motives of their own.

As we move from anticipation of the discovery of DNA in the 1940s to anticipation of the completion of the Human Genome Project in the twenty-first century, these metaphors tend to accrue rather than replace one another. Figurations move to increasingly narrativized analogies such as the book of life or the world's greatest history book, which imply sweeping answers to the mysteries of life. The figures of the Holy Grail and the Rosetta stone evoked in relation to the Human Genome Project hyperbolically combine the implications of all other analogies with their own cachet of legendary importance. In a final reduction of all of this to a "parts list" after the completion of the Human Genome Project, Eric Lander, head of the Whitehead Institute at MIT, offers a potentially corrective humility that hopes nonetheless to perpetuate public interest and funding by evoking a different aspect of the same productive agenic narrative imported by other textual analogies.[8]

The layering of figuration is not unusual insofar as any scientific analysis requires this set of representational transmogrifications for nonspecialist dissemination. The thing is never the thing the moment we elicit comparisons, muster narratives, and via those narratives bootstrap imaginary capabilities and unacknowledged values. That representation is inevitably misrepresentation is an inherent feature of language, which, taken to its logical end, suggests that no scientific phenomenon can ever be represented without some kind of distortion. The issue is not so much the inevitably distorting effects of language on the "truth" of science but what happens when descriptive analogies take on a life of their own, hijacking basic principles and operations, adding values, superimposing more convenient, compensatory, or ideologically felicitous narratives that ultimately supplant more accurate (or less deviated) characterizations.

The DNA gene's own representational slide begins with the etymology of the word *gene* itself, which reflects the concept of the gene that Mendel implied in his 1865 paper "Experiments in Plant Hybridization."[9] Mendel refers to the agents that carry inheritable "characters"—or traits—as "elements." To Mendel these elements were material and were carried from one generation to another. Whether Mendel meant "element" in simple structural terms or as connoting something more like the atomic elements of the periodic chart, it is clear that he meant something basic and fundamental.

Mendelian ideas of heredity were more or less ignored until they were "rediscovered" around the turn of the twentieth century. William Bateson, the translator of "Experiments in Plant Hybridization," reintroduced Mendel's ideas to English-speaking audiences and himself proposed experimental work in genetics.[10] In 1906 in a speech to the Third Conference on Hybridisation, Bateson introduced the term *genetics:* "I suggest . . . the term Genetics, which sufficiently indicates that our labours are devoted to the elucidation of the phenomena of heredity and variation."[11] The first term coined refers to the overall field of heredity and not the specific "elements" by which "characters" are passed from generation to generation. Genetics encompasses the effects that could be measured through experiments and only points, as did Mendel's findings, indirectly to some supposed element at the heart of the unknown processes that produced these results. The term *genetics* also implies that the field of heredity is the study of generational reproduction as a way to explain both variation and survival, individual differences and statistical patterns of similarity.

This newly named field of genetics then itself spawned the term *gene,* coined by Wilhelm Johannsen in 1909. In the *American Naturalist* in 1911, he notes that he "proposed the terms 'gene' and 'genotype' . . . to be used in the science of genetics. The 'gene' is nothing but a very applicable little word, easily combined with others, and hence it may be useful as an expression for the 'unit-factors,' 'elements' or 'allelomorphs' in the gametes, demonstrated by modern Mendelian researches."[12] Johannsen also adopted other vocabulary to name the implied phenomena of heredity—"genotype" as an organism's actual genes, "phenotype" as the organism's displayed characters—filling out a conceptual picture of the relations of observed and implied phenomena involved in transmitting characteristics from generation to generation.

This little matter of vocabulary is more important than it might seem, as it provides the first instance, at least in relation to the gene, of how the use of language adds concepts and values to basic biology. Just as genes constitute the material elements of genetics, so the word *gene* is a little constitutive element of the term *genetics* and was prophetically invented or discerned in the same way that genes themselves would actually be discovered as an effect of studying heredity. But, it is also claimed, Bateson's term *genetics* did not gain popularity until Johannsen proffered the term *gene*. This claim, whether true or false, reorders words and concepts so that "genetics" appears to come from "gene" instead of the other way around, as in the order of their historical appearance. This reordering exemplifies not simply the inevitable and perhaps unresolvable mismatch between any process and the way it is represented, but also that there is something more at stake in the apparent vagaries and associations imported by such incidental misrepresentations. In this case, if "gene" is derived from "genetics," then the field of heredity focused on the transmission of similarity, and variation, is more than the operation of its atomistic part. If "genetics" owes its acceptance to the appearance of "gene," then the gene becomes the causal key of everything and the originary element that spawns everything else: words such as *genotype, genome,* and even, falsely, *genetics* itself.

If this minor reversal weren't a sufficiently telling symptom of how language and analogies inevitably reflect and import certain investments, later sources provide another etymology (and story) for the origin of the word *gene*. Some dictionaries (such as *Merriam-Webster's*) credit Johannsen's choice as deriving from the German word *Pangen,* a noun referring to the particles formerly believed to have been contributed to the reproductive cells by various parts of the body in a process called "pangenesis." If *gene* is indeed a short form of *pangenesis* rather than simply a post facto root borrowed from *genetics,* then the coinage of *gene* stands for a correction of the mistaken hypothesis of pangenesis, eliminating *pan* and the *sis* as the key originary element but carrying with its piece of earlier terminology a history of theories about it (something the chromosomes will also be credited with). Other dictionaries avoid the German detour through pangenesis altogether, simply linking the word *gene* to variants of the Greek *genos,* meaning race or offspring. This last etymology privileges the generative (also from *genos*) nature of genes, linking them simultaneously to reproduction and to the traditional Greek locus of the origins of Western culture.

Although the etymology of a word and the history of its appearance ask two different questions, the ways the term *gene* actually takes over the fields of heredity and reproduction anticipate both the central position the concept of the gene will come to have in future research and how ideas about genes are crucially altered and augmented by the very language employed to describe and understand them. There is, in other words, something inherently linguistic and performative about our idea of the gene reflected in its terminology from the start. From the moment of its appearance, the term *gene* was believed to generate other words, even as it comes from the root word for generation. Linguistic generation created a portrait of a generative process that echoed and inflected how we think about genes. If words come through a derivation from an old word (*genos*) and if, through the transmission of a syllable ("gen-"), genes produce a series of variations (gene, genotype, genetics, generation), isn't that how genes themselves might work? The concept meets the analogy before most researches ever even get started.

The idea of the gene as etymologically foundational projects an alternate story onto genetic research in the first half of the twentieth century. Historically, rather than being the bit that starts it all, the gene is a part that must be discovered and characterized. The idea of the gene occurs first by implication in the effects of hybridization, then through statistical strategies of mapping chromosomes, and then through the deductions of molecular biology that come to identify DNA with the concept of the "gene" but in so doing must also break the concept down into a number of processes.[13]

For another half century, the word *gene* stood only for a function—the transmission of "characters." What is surprising is the extent to which the details of the gene's functions were anticipated and understood while the actual mechanisms and molecules responsible remained a mystery (even though the chemical substance DNA had been known since 1869). The analogies and figurations of DNA as the genetic agent that prefigure and anticipate the working out of its structure already determine (or overdetermine) the directions research takes as well as predefine the analogies and coincident conceptual baggage that accompanies DNA as the mode of its public transmission. The specific trajectory and choice of DNA analogies trigger a series of narratives that import a surplus of assumptions about agency, control, and the order of things that constantly displace a complex systemic way of thinking with the familiar structures, shapes, and (re)productive impetus of Western narrative. The trajectory from code (instead of cipher) to template to performative alphabet leads from the systemically mechanistic to the predictably referential and ineffably structural.

Persistent Preemption

This trajectory begins with Erwin Schrödinger's suggestion that nuclear structures called chromosomes carry what he calls "a code-script," which for him sometimes functioned as the seed of narrative and other times became an elemental instance of performativity, taking on the avatar of the Morse code. "In calling the structure of the chromosome fibers a code-script," he says, "we mean that the all-penetrating mind, once conceived by Laplace, to which every causal connection lay immediately open, could tell from their structure whether the egg would develop, under suitable conditions, into a black cock or into a speckled hen."[14] The gene is an instant story, like a sponge that when dipped in water expands into the shape of an animal. The ultimate form of this sponge is divined by an "all-penetrating mind" that through a kind of reading or analysis tracks causal connections as they unfold. This presumably is the "script" part of the code, a dynamic ordering that makes sense of the code itself. Schrödinger's analogy already imparts the meaning (here gender) coincident to the narratives of (re)production hovering alongside evocations of structure and generation—here evident in his gratuitous examples, especially in the ways it worries both about the transmission of characteristics as such ("black cock" or "speckled hen") and about how those characteristics unfold in the development of the species.

For Schrödinger, thus, genes are multifaceted: "They are law-code and executive power—or, to use another simile, they are architect's plan and builder's craft—in one."[15] Reaching for the kind of performativity formulated by Austin in the speech act, Schrödinger predicts that structure is function—that what the "code" is, is also the means by which what it says becomes an organic reality. Just as an architect's plan shows how a building is put together, so Schrödinger predicts, genes will show both the materials of an organism and their arrangement through time (in development) and space (in the species' form). But the plan does not in itself build the house. Schrödinger's claim is that the gene does that as well and as an effect of the same molecular arrangement as the plan. Plan and craft merge as one. Schrödinger's code-script is already plenipotentiary.

The combined analogies of code-script and plan-craft don't tell us much about how the chemistry of chromosomes actually manages to fulfill these functions. For this, Schrödinger turns to the analogy of Morse code as a model for how molecules might communicate. "Indeed," he comments, "the number of atoms in such a structure need not be very large to produce an almost unlimited number of possible arrangements. For illustration, think of the Morse code. The two different signs of dot and dash in well-ordered groups of not more than four allow thirty different specifications."[16] The code analogy explains how a small number of molecular variations might produce a sufficient number of different signs, as the Morse code's arrangements of dots and dashes permit the translation of the alphabet. But because in addition, as Schrödinger insists, it "must itself be the operative factor bringing about the development," it is also a material line of molecules whose arrangement in itself corresponds "with a highly complicated and specified plan of development and should somehow contain the means to put it into operation."[17]

Schrödinger's predictions, made at the end of World War II, were in a curious way like genes themselves. On the one hand, they specified with uncanny accuracy not only that genes are arrangements of molecules that are themselves the information such arrangements bear, but that the arrangements themselves are the means by which information moves from generation to generation. On the other hand, Schrödinger's ideas themselves became the catalyst for the discovery of DNA. Watson recalls that, as an undergraduate at the University of Chicago in 1946, he had suddenly become aware of the gene through Schrödinger's book *What Is Life?* "Soon after its publication in the States, I spotted this slim book in the Biology Library, and upon reading it was never the same."[18]

Given the influence of Schrödinger's ideas on the youthful enthusiast Watson, it is not surprising that the notion of the code stuck as a premier analogy for understanding the role of chromosomal chemistry. But with Watson and Crick's attempts to describe the structure and significance of DNA, the code analogy oozes quickly into the additional (and significantly different) analogies of language/text and template. In their second article about DNA, Watson and Crick liken the structure of DNA to a code: "It follows that in a long molecule many different permutations are possible, and it therefore seems likely that the precise sequence of the bases is the code which carries the genetical information."[19] This succeeds an earlier allusion Crick makes to code in a letter to his son written in March 1953 (one month before publication of their first letter about the structure of DNA in *Nature*), where Crick describes the pairing of the nucleic acid bases as a "code. If you are given one set of letters, you can write down the others. Now we believe that D.N.A. is a code. That is, the order of the bases (the letters) makes one gene different from another gene (just as one page of print is different from another)."[20]

These notions of language/text and code also quickly merged with the idea of a template. In their second 1953 paper on DNA, "Genetical Implications of the Structure of Deoxyribonucleic Acid," Watson and Crick speculate that DNA is a "template" at least as far as the mechanism of its self-reproduction goes: "We feel that our proposed structure for deoxyribonucleic acid may help to solve one of the fundamental biological problems—the molecular basis for the template needed for genetic replication. The hypothesis we are suggesting is that the template is the pattern of bases formed by one chain of the deoxyribonucleic acids and that the gene contains a complementary pair of such templates."[21]

The positive influence of metaphors certainly accounted for the initial directions of molecular biology theory, notably the direction adopted by the physicist George Gamow. Gamow took both the template and the code ideas literally, hypothesizing that DNA molecules synthesized proteins by acting as a physical template for amino acids. He proposed that the four bases (A, T, C, G) constituted a "four-digital system" (i.e., four integers instead of ten). Listed in the order of their appearance, these digits produced a long number—he called it "the number of the beast"—that would characterize members of any given species.[22] The order of the bases, according to Gamow, would produce small diamond-shaped holes into which would fit, according to their shapes, "various amino acids . . . as specifically as keys into locks."[23] Held in proximity by this template DNA chain, the amino acids would bond with one another to form proteins. Gamow was further inspired by the fact that the geometry (and chemistry) of diamond-shaped chemical holes meant that there were only twenty possible combinations of bases, a fact that corresponded with the number of amino acids thought to be active in the synthesis of proteins.

While Gamow's mechanistic hypothesis quickly proved false because of a combination of less-than-informed chemistry and his ignorance of the role of RNA, his idea that the sequence of bases somehow refers to the manufacture of amino acids was more fruitful. Thinking of DNA as a chain of bases whose sequence somehow orders molecules into amino acids makes the idea of a code even more attractive. And, with the rejection of the template idea, the idea of the code requires even more analogies to conceptualize how a sequence of bases can drive the manufacture of anything other than itself. On the one hand, as Horace Freeland Judson observes, this results surprisingly in "the perception of the biological code as an abstract problem distinguishable from the biochemistry."[24] On the other hand, the abstraction of the code invites a plethora of linguistic and even textual metaphors linked to conceptions of coding. This is as opposed to Schrödinger's earlier suggestion that the code-script works like Morse code, which is a cipher.

The small distinction between code and cipher is symptomatic of the agenic structural direction of analogies and their implied narratives that emerge as the representational apparatus surrounding DNA and genes after Watson and Crick. The analogy of the code imports whole notions of language and texts, while the analogy of the cipher, which refers quite literally to nothing (to say something is a cipher is to say that it stands for nothing, from the Arabic *sifr,* "zero") imports only mysterious notions of international espionage. In a code, one symbol stands for another symbol, concept, or condition arbitrarily but consistently. The alphabet, for example, is a code in which letters stand for phonetic sounds (usually). There is no rhyme, reason, or relation that would account for why a particular written shape should stand for a particular phonetic sound: the relation is arbitrary (despite, for example, imagined correspondences between the shape of the mouth and the letter *o*).

Language is a code in the same way. A cipher is a way to encode (or encrypt) one set of symbols in terms of another, where the encoding is performed according to a set of rules. For example, if we shift the alphabet five letters to the right, the word *gene* becomes *ljsj*. If we know the rule, then we can read the cipher by counting back five letters. When the set of rules is applied to the encrypted message, it can be translated back to its original form. The relation between the encryption or cipher and the text it encodes is not arbitrary, operating as it does according to a strict, if often complicated, set of algorithms (procedures or patterns for translation).

To compare DNA's operations to a code is to suggest, perhaps unwittingly, that the relation between base sequences and what they may catalyze or produce is arbitrary. To think of DNA as a cipher would mean that DNA's processes work according to a nonarbitrary set of discernible rules. Thinking about this difference later, Crick admitted that perhaps Schrödinger's analogy of Morse code, which they all knew about, defined the direction of their thinking, but pointed out that even the term *Morse code* should be *Morse cipher:* "You understand," he comments, "I didn't know that difference at the time. 'Code' sounds better, too. 'Genetic cipher' doesn't sound anything like as impressive."[25]

Code it became and the analogy took over, defining instead of merely describing. In his early essay expounding his template/code theory, Gamow, for example, says, "If one assigns a letter of the alphabet to each amino acid, each protein (and, in particular, each enzyme) can be considered as a long word based on an alphabet with 20 (or somewhat more) different letters."[26] The nonarbitrary relation between base sequence and product has already evaporated in favor of an idea of signification—of the fact that sequences might mean something in a regular and predictable fashion. This meaning quickly comes to be understood in terms of abstractions—the ultimate product, heredity, eye color—

instead of in terms of the immediate biochemical products of activity around and with DNA.

So if, for some reason, the concept of the code snuck in as an attractive analogy in the incipient discoveries of DNA genes instead of other perhaps more apt comparisons, what, other than its impressiveness, does the idea of the code import to the idea of the DNA gene? How does the analogy of the code join with the earliest confusions about the connection between concept and word present in the invention of the word *gene?* How does the relation between *gene* as already the word that spawns words and other linguistic and textual analogies alter and evolve public ideas of DNA genes, their value, and their possibilities? One answer is that the code/language analogies privilege ideas of reproduction, manipulation, and survival. The word *gene* is understood, at least in one account of its etymology, to generate other words; the basic meaning of its Greek etymology is generation. Textual analogies offered to explain DNA import an uncharacteristic mechanism for how it may be that DNA works, which enables certain kinds of powers (authorship, alteration, ownership) while allaying certain fears (loss of control, extinction, insignificance). Scientific concepts may always be misrepresented. What is important is how those adjustments take on a life of their own.

A, B, C Unlimited

The prevalence of book and language metaphors for DNA and genes is so great that it would be impossible to list them all. The genome as the book of life has been the most prominent of these, and we usually take the idea of DNA as an alphabet for granted. While geneticists and molecular biologists may have abandoned linguistic analogies long ago as ways to understand their own research, there are any number of popular commentaries that present linguistic and textual analogies enthusiastically. Jeremy Campbell's 1983 *Grammatical Man,* Steve Jones's 1993 *The Language of the Genes,* Walter Bodmer and Robin McKie's 1994 *The Book of Man,* and Nicholas Wade's 2002 *Life Script: How the Human Genome Discoveries Will Transform Medicine and Enhance Your Health,* for example, anticipate the torrent of appeals to the alphabet and book precipitated by the announcement of the genome's completion.[27] On June 26, 2000, John Toy, medical director of the Imperial Cancer Research Fund in Britain, announced, "We have discovered the human alphabet—what we now have to do is put the letters in the right order and make a sentence. Only when all that is done shall we have the book of life to read." On the same occasion, Watson declared, "The human genome is our genetic instructions and it's as if we have a series of books that provide instructions for human life."[28] "In June," Lander comments, "maybe people thought we had this big pile of letters and it was all stuff. But I don't know if people realize that we just found the world's greatest history book. We are going to be up every night reading tales from the genome."[29]

If the notion of the book, language, or text does not frame discussions of DNA genes, most popularizing authors employ linguistic and textual metaphors as the primary figure to explain genes, DNA, and the genome. Matt Ridley's *Genome: The Autobiography of a Species in Twenty-three Chapters* is an example of the extent to which textual metaphors define popular notions of DNA's operations.[30] A "lucid and exhilarating romp through our 23 chromosomes" (as Watson comments on the book's back cover), *Genome* depends entirely on understanding genes (and DNA) as a kind of script or codex of instructions for human beings, as a transcript of the past, and as software. Focused on the genome, or the entire list of base nucleotide pairs from all twenty-three human chromosomes, Ridley's sampling of meaningful sites on each information-packed strand transforms human genes into an index of individual traits ("Intelligence," "Instinct," "Self-interest"), social concerns ("Politics," "Eugenics," "Environment"), and life themes ("Life," "Fate," "Death") that constitute both the physics and metaphysics of twentieth-century Western humanity. The genome is a book, Ridley explains, with "twenty-three chapters, called chromosomes," where each chromosome contains "several thousand stories, called genes," where "each story is made up of paragraphs, called Exons, which are interrupted by advertisements called Introns," and every paragraph "is made up of words, called codons," and "each word is written in letters called Bases."[31]

Richard Dawkins, one of Ridley's precursors and more particular in his usage than Ridley, sets out his scheme of analogies in *The Selfish Gene:*

"I shall make use of the metaphor of the architect's plans, freely mixing the language of the metaphor with the language of the real thing. 'Volume' will be used interchangeably with chromosome. 'Page' will provisionally be used interchangeably with gene, although the division between genes is less clear-cut than the division between pages of a book. This metaphor will take us quite a long way."[32]

Both Ridley and Dawkins—as well as most other commentators (including scientists themselves)—succumb to the irresistibility of mixing these metaphors. Dawkins combines "architect's plan" with the figure of the book. Ridley mingles books, software, narrative (or story), and history. Part of this mixing comes from DNA's multiple roles. Part, however, enacts exactly the ways analogies slide from one suggestion to another, based both on their different uses in relation to DNA genes and on their suggested affinities to one another. At some point analogy generates more analogy. Mixing analogies with "the real thing," as Dawkins warns us he is doing, has the effect of "real"-izing the analogies—of making analogy indistinguishable from more accurately descriptive terms. Leaping from the "architect's plan" to a series of analogies of the book shifts figures in midsentence, suggesting that the book supplements and elaborates "the plan"—as if there were a blueprint and extensive written commentary. The relations among metaphors, as both examples illustrate, are typically hierarchical. This hierarchization reveals a way of thinking about DNA genes as building blocks of meaning and as constituents of a meaningful organization that tends toward something grander—a history, a plan, even a metaphysics that renders DNA not only the "secret of life" but also the secret of Life.

If DNA genes work like a language, read like a book, and spell out the fate of humanity, how exactly does that mean that they operate? What are the processes for which linguistic metaphors provide an analogy? Language consists of a system of "signifiers"—words that refer to things or ideas, called "signifieds." There is no essential relation between a word and the thing to which it refers. Only convention moors the connection between signifier and signified. English speakers agree, for example, that the word *tree* refers to a large woody plant, and French speakers agree that the word *arbre* performs the same function. Words, thus, work as symbols. On the level of discourse—of the sea of language and ideas within which we all exist—words substitute for the things to which they refer. The process of language and by extension any of its parts and products—the alphabet, the sentence, the book—is substitutive.

The symbols that constitute language stand for a thing by means of a leap, since there is only an arbitrary and not an essential relation between them. This arbitrary relation and process of substitution means that the operation of languages and texts is not at all mechanical—is not defined or driven by physical necessity. Words stand for things and may stand for one another. A variety of words may be chosen for the same thing or idea—tree, maple, oak, pine, woods. Although convention may seem to demand it, no word necessarily requires another, and no idea necessitates a specific word. This means that language permits manipulation, choice, authorship, and originality. We can always make new combinations of words (think of how advertising employs this idea), but DNA cannot make any but the combinations already defined by the physical attributes of its parts.

As metaphors on behalf of metaphor—as metaphors that suggest that DNA works through the same substitutive logic as language where one word or thing arbitrarily stands for another—linguistic and textual metaphors suggest a mode of DNA reproduction that occurs through substitution and combination in the face of the very different kind of logic of duplication at work in genetic chemistry. We know genes are involved both in their own reproduction and in the reproduction of organisms, but DNA's mode of replicating is in fact different from the conjunctive bliss we normally associate with the egg-and-sperm saga (whose "courtship" behaviors are often romanticized in the hackneyed epic of the stalwart chevalier and the reluctant bride). To repeat the simple version of DNA self-replication: DNA is a molecule composed of two attached chains of linked nucleic acids. The four different nucleic acids come in two complementary pairs of "bases"—adenine (A) and thymine (T), and cytosine (C) and guanine (G). These bases are paired like rungs on a ladder in a chain bound together with sugars to produce DNA's famous double helix structure. If the ladder is split apart in the middle of each rung, each side provides an empty set of slots that can be filled only with the complementary opposite of the pair. If, for example, one side of a DNA ladder consists of A, C, G, T, it can be reconstructed only by reinserting the complementary pairs T, G, C, A. When the base pairs "unzip" and replicate, two identical ladders of the series are produced. In vitro DNA replicates itself with the help of another nucleic acid, RNA, which forms a "negative" impression of the base pairs by matching the bases with their complementary opposites (except for thymine, which is replaced on RNA with another base, uracil).

DNA replication, then, though it recalls romantic myths through its literal attraction of opposites (which are "opposite" only because of their structural position as complementary parts of pairs), works through mechanical contiguity—the alignment of one molecule with another specific molecule—cytosine with guanine, thymine with adenine. Because of the shapes of the nucleic acid molecules, only certain specific acids will form pairs with others. This pairing is defined by the literal shape and charge—the chemistry—of the molecules. The process of DNA replication is thus metonymic (based on the relation of elements side by side) instead of metaphoric (based on the substitution of one like element for another) in its logic. It is like a photographic negative. Its series of base pairs, when replicated by RNA, produce the proteins that eventually constitute living flesh—again, more like a photographic negative, a template, or even molecular topography than a word. Individual genes, which might be composed of thousands of base pairs, contain not only amino acid templates for proteins but also DNA that initiates or closes down certain processes, that provides a protective margin of extra base pairs, or that may work to produce several different enzymes by simply starting at a different place. DNA and genes in concert with enzymes produce only nucleic acids, amino acids, proteins, and the initiation or termination of replication through a thoroughly mechanical process. This process is not, in fact, even quite digital, in that it has two binaries instead of only one, and those binaries are complementary (they attach to one another) instead of oppositional (where they substitute for or replace one another as do the off/on, 1/0 binaries in computers. DNA is, thus, mechanical with none of the arbitrariness or indeterminacy of language: C always requires G, for example). There is no gap of arbitrariness or choice to be filled between base and base, as there is between signifier and signified. In contrast, language is substitutive, arbitrary, and manipulable.

The difference between how DNA operates and how languages and texts work recalls an important distinction discerned by both psychoanalysis and linguistics between two basic facets of human thought. Let us assume that Sigmund Freud and Roman Jakobson were right when they each proposed that only two basic logics (or "poles," to quote Jakobson) order language and representation.[33] On the one hand, condensation or metaphor enacts a logic of substitution and combination that typifies the operation of language itself as it replaces things, actions, or ideas with words and their combinations. Metaphor represents an interplay of similarity and difference when a different signifier with some similarities to another signifier stands in for it. Metaphor produces a pyramidal effect on a vertical scale (to speak metaphorically)—that is, one word stands for another, which stands for another. On the other hand, representation can also be seen as a process of displacement or metonymy, as one word relates to another by being next to it in a chain such as a sentence. Metonymy works through contiguity—by being next to or associated physically with another word. Metonymy produces a string or chain effect on a horizontal scale (again to speak metaphorically).

These logics of representation, while initially elicited to explain the workings of dreams and language, also define the difference between the magical or symbolic (including representation itself as a mode of conjuring those things that are not present) and the physical mechanical world, which operates quite literally through the meshed contiguity of parts. Another way to understand the difference (and the relation) between metaphor and metonymy as organizing logics is represented in the traditional differences between father and mother as well as other such binaries as soul and matter. Before DNA technologies, a father's relation to his child was always metaphoric; the patronym cemented a relation that could never be entirely certain. In other words, the essential uncertainty of a father's relation to his child was sutured over by using the father's name. The name substitutes and signifies a relation that, until DNA testing, was never really quite certain. In contrast, a mother's relation to the child is always metonymic; she is a child's mother by virtue of the physical contiguities of birth. The metaphor or linguistic symbolism of paternity is a type of magic—of making a relation where none may exist. This symbolic magic typifies law in general, which also uses language to produce relations that otherwise do not exist (such as between a piece of land and an owner). These symbolic or metaphoric functions are connected to soul or spirit precisely because they escape physical determination and instead occupy (and in fact produce) a realm where language can effect deeds. Mother, or mater, with her metonymy inhabits the realm of matter.

The politics of sexual difference that haunt the parental scenario also define the relative value of metaphoric and metonymic logics. The symbolic and magical is generally valued more than the mechanical and the literal; the incarnate is worth more than the material and embodied. When the mechanical is valued (as it often is), it is metaphorized; hence the speeding train becomes a symbol of human progress or the computer becomes a "brain." In other words, metonymy, though its logic rules technology and even the workings of biology, will only slowly come to dominate representational logic and symbolic systems, and only if it is consistently veiled by or transformed into metaphor. The logic of substitution (symbolization, paternity, soul, magic) eclipses the contiguity that underwrites it, even—or especially—in relation to DNA, whose operation is purely metonymic. In this sense, DNA, genes, and metonymy are a figurative mother whose regime is thwarted and constantly reshaped by the representational machinations of an increasingly obsolete paternal law.

Through their linguistic analogy, DNA's textual metaphors enable illusions of both genetic literacy and propriety, providing models for ideas of authorship and ownership that have attached to DNA and genes since the 1980s. If DNA is a complex text to be deciphered, science becomes art, scientists become paleographers and critics, and the workings of vital chemistry can be authored, copied, rewritten, and copyrighted by anyone who can wield the "language." It doesn't make any difference that DNA chemistry really can be reworked, though in ways that barely resemble literary criticism or textual analysis. Rather than provide a key to chemical methodologies, textual metaphors support the right to decipher, decode, rewrite, own, and profit from life's "textual" key.

DNA's textual analogy also implies such other genres as narrative and history and such activities as transcription, translation, and reproduction, which also become part of what we envision as genetic function and process. If genes are a text, the text is as much history as recipe, and the evolution of humanity can be "read" in its figurative etymologies, archaic vocabularies, and "verbal" excess. DNA's extensive "book" contains the vocabularies of both earlier forms of humanity and genes that appear in other species. Like any text, genes can represent a range of meanings without being alienated or diverted from their human narrative. For example, humans share 98 percent of their genes with chimpanzees. But if archival remains and textual wanderings of genes are not themselves accounted for (as in the Human Genome Project), the sense of a unified set of specifically human genes may never be established. Since the human genotype is mostly not unique to humanity, humans deploy other modes of appropriation such as naming genes, listing and mapping them, and ultimately producing and patenting them. Instead of a Frankensteinian overreaching, this is ultimately a flustered attempt to mark the code with the ineffably human in the face of fears that DNA extends beyond both species and era.

Hanging on the Paternal Prerogative: The Reproductive Narrative

What are the effects of or perhaps even motivations for this metaphorization? First, the metaphor of language helps perpetuate traditional narratives of reproduction. These reproductive narratives perpetuate fading systems of the patriarchal family, binary ideologies (such as them/us or male/female), and economic systems premised on the bifurcation of roles and possibilities (such as class). Second, linguistic and textual metaphors underwrite the continued viability of concepts of individuality in an environment of ambivalence partly caused by DNA's somewhat leveling potential (we all have the same genes. We have most of the same genes as chimpanzees. Genes are the causes of behavior, not human will). Third, textual metaphors suggest that genes can be manipulated, rewritten, and ultimately owned. Patenting genes and turning them into property provides a powerful motive for thinking of genes as texts. Finally, metaphorizing DNA as language perpetuates systems of patriarchal law that have been threatened by technological change. For example, in undermining the necessity for the paternal metaphor (the use of the father's name) by proving the connection between father and child, DNA also threatens the collapse of the symbolic order that depends on the function of metaphor as the bridge between matter and culture.[34] The fact of paternity is not as important as the means by which it is established and sustained—metaphor, law—are. To avoid the collapse of paternity and law, DNA is rewrapped in metaphor—and not just any metaphor, but the analogy of the Word, the code, the same figure applied to originary paternity and law (which, at least biblically, came together).

In keeping with their etymology, genes seem to fulfill the promise of the long-sought single mechanism that would provide life's key, since genes, like life, can instill order and reproduce themselves. Perhaps not uncoincidentally, order and reproducibility are also traits we associate with language. Textuality and life would seem to have some common ground, the latter easily translated into the former. But genes go one step further than texts, which is what makes genes alive: DNA is credited with being able to reproduce both life and itself. The gene—the generative element—is the quintessential node of reproduction; it reproduces both itself and us. By reproducing itself as well as the chemistry for its own reproduction, the "secret" of life accounts for both the reproduction of order (that complex systems are deployed in a working arrangement) and the orderly reproduction of individuals within a species.

But even though DNA is primarily employed to produce and to maintain complex systems, the attractiveness of the gene seems to consist mainly in its role as a purveyor of heritability. Discussions of genes emphasize their role in human (or other species') reproduction; genes bring the past forward as something both same and different. This notion of reproduction as same but different is opposed both to production, which produces multiples whose form differs from the sources and processes of manufacture (a tractor is different from both its constituent elements and a tractor factory), and replication, which comprehends only the production of the self-same (photocopying). Since DNA both produces and replicates while it reproduces, our fixation specifically on the genes of individuals as a link to the past points to a preoccupation with our own individual origins.

We might, then, understand the question that DNA genes answer as the question "Where do I come from?" The DNA gene provides an answer in the realm of the minute, which while sharing all of the cachet of molecular rigor, turns out only to defer the question of the origin of the molecule itself and that origin to the origin of the planet, and that to the origin of the universe, and so forth. As it offers a specific unique cause for an individual and the mechanism for the production of more, the DNA gene also extends the range of origins from local parentage to epic (and even epoch-exceeding) history to the point where origins are meaningless—to the point where cosmos and DNA meld. DNA threatens to remove human reproduction from the realm of the culturally sacred to the environs of cosmic happenstance. For this reason DNA itself, although historically ambivalent, is all the more insistently appended to the large-scale reproductive incident—that is, is located as the agent of specific identifiable passages of heredity. Thinking of DNA primarily as an agent of heredity—as the biological confirmation of a notion of kinship—comes at the expense of the myriad other processes DNA genes manage that have nothing at all to do with reproduction.

We might claim that this bias toward reproducing individuals occurs because of the gene's nature, or the state of evolving scientific knowledge, since heredity and the effects of mutations were the first evidence of genetic function. But our focus on reproduction might also be because figures of reproduction intrinsic to notions of textuality are particularly attractive and necessary to our cultural acceptance of the idea of the gene. Not only does reproduction center the human in the genetic process (as opposed to centering the gene whose interests, as Dawkins suggests, are sometimes different from our own), but we understand reproduction as working like a metaphor—itself a collision of different and same that results in a new meaning. In reproduction, as in texts, metaphor is the logic through which all meaning is produced. Retaining the gloss of good old-fashioned reproduction means that even after the discovery of DNA genes, we can still think of both meaning and (re)production in the same ways as before their discovery. The metaphor of the text enables a painless transition through beliefs about reproduction that avoids as much epistemological change as possible. In this way old systems such as patriarchy continue to dominate the ways we conceive of the world, even in the face of mounting evidence otherwise, even when, if we can manipulate the very processes of genetic reproduction, the notion of a father may become irrelevant.

The combination of textual metaphors and narratives of reproduction comes back around again. DNA's textual analogy as the analogy of the text itself becomes increasingly literalized. Genetic behavior appears as human reproductive behavior in small and follows the reproductive aegis of Western narrative itself, operating in a manner similar to metaphor. Two somewhat dissimilar entities—say, body and text or man and woman—come together and produce through their combination a new meaning (property, child) that elaborates and might even substitute for one or both of the original entities, is similar to them, but continues as itself. This narrative of reproduction is the literalized, material figuration (if any such oxymoron is possible) of metaphor, epitomizing its processes of substitution, condensation, and signification. Imagined as occurring on a vertical plane, human reproduction is represented on the branches of a family tree. The verticality and the layered substitutions of metaphor produce history, tradition, narrative, and order in the guise of the law that functions, like the gene, to impose order on an otherwise chaotic body politic.

Both deploying metaphors of textuality and emphasizing the genes' role in reproduction turn the genes' process of replication (the production of exact copies) into the more picturesque vicissitudes of reproduction. Genetic replication is metonymic and automatonic in the sense that it occurs automatically and autonomically based only on the nucleotides present. Glossing this process with visions of sugarplum chromosomes dancing together to mingle their codes coats genetic mechanisms with the combinatory romance of metaphor. While certainly during conception chromosomes do meet, match up, and trade their genes, and that matching affects the operation of specific nucleotides present, the chromosomes continue to replicate rather than reproduce—they copy only what they have rather than make any further new meaning (until they age or are smitten with toxic substances, and then their new meaning is often cancer). The point is that the persistent use of metaphor draws attention to romantic notions of reproduction that then become the terrain within which genes and genetics operate. This in turn romanticizes the gene, which becomes an attractive little personal representative on the level of the molecular. This little representative superimposes the narrative of romantic heterosexuality onto chemical processes so that genes are little more than the same old story in miniature.

If we imagine that genes operate like little romantic couples, such a metaphor obviously introduces an element of binary heterosexualized gender as the very basis of genetic operations, a phenomenon treated at more length in the next chapter. It also suggests genetic agency. If we imagine that genes have agency, that they act like little homunculi or men who, for their own reasons, make one choice or another, then we easily transfer entire sets of human characteristics, desires, motives, and idiosyncrasies onto the more or less automatic operation of genetic chemistry. This substitutes a logic of human psychology onto molecular biology and renders the entire genetic process less the by-product of chemical operations (that would happen with or without us) and more a mirror of our own will. In representations of genes for the general public, this figure of genetic agency plays out in two somewhat opposing and paradoxical ways. On the one hand, genes are very much the object of human control. If DNA is a text that we can read and if it is something we can transcribe, then it is also something we can copy, selectively manufacture, and alter. Instead of being its subject, DNA becomes ours; humanity stays unique, and the higher powers of mystery and magic still reign. This enables our continued belief in the powers of both humanity and the individual rather perversely in the face of discoveries about the role of genetics in behavior, the idea of the selfish gene (or that genes are in fact out for themselves), and the utter complexity of genetic operations.

Individual Angst

On the other hand, and paradoxically, we would like to see genes as agents that control human behaviors. Whether genes might really produce certain behaviors is actually irrelevant. What is important is that in imagining a genetic cause for all behavioral phenomena—and particularly aberrant or deviant behaviors such as criminality, attention-deficit disorders, homosexuality, and antisocial behaviors—we transfer responsibility for certain kinds of problems from the realm of the social to the genome, both bypassing the individual as any source of expected control and yet emphasizing the individual as the bearer of idiosyncratic genes. This actually represents an ambivalence about the status of individuality. We are in control of our genome but not of our behavior. Our genomes are unique, and yet there is a human genome that is treated as if it is the genotype for everyone. Our DNA identifies us, yet we share 98 percent with chimpanzees (and how much with pigs, snakes, cockroaches, pumpkins, or plankton?).

Loss of individuality is a source of anxiety in the face of genetic discovery, especially in an era when the illusion of individualism sustains and is created by participation in consumer culture. We both reflect and allay that anxiety by foregrounding DNA's forensic uses in an ever-increasing number of television programs centered on molecular evidence: *CSI, CSI: Miami, CSI: New York, Forensic Files, The New Detectives,* and so forth. We emphasize individual choice in realms where it doesn't matter, such as clothing style; give people names and social security numbers to signify both continuity and uniqueness; and practice sales surveillance techniques that deploy the myth of the individual as the centerpiece of marketing campaigns (the best example was one's name writ large on Publishers Clearing House contest entries). Within families, and currently in the culture at large, we interpret links between heredity and behaviors as individualized quirks that simultaneously distinguish and exculpate individuals.

From the discoveries of Seymour Benzer and the drosophilists to more contemporary genetic researchers, it has become clear that genes may govern more than physiology.[35] Determining to some degree some kinds of behaviors (actually an ever-increasing list), knowledge of genetic influence or causality increasingly shifts responsibility for such behaviors as anger, hyperactivity, drug addiction, intelligence, sloth, and thrill seeking from the individual to the genotype. If our genes comprise a governing life text, then the role of individual choice seems to diminish.

And if, as Ridley points out, genetic predispositions may influence behaviors from 40 to 60 percent of the time, then that influence offers the same percentages to which our own responsibility for behaviors is diminished. At its most extreme, this genetic causality becomes a determinant beyond the individual's control, shifting notions of culpability and responsibility and suggesting such possibilities as the "my genes made me do it" defense to criminal behavior. It has become a commonplace to attribute behavioral disorders ("lying, stealing, bullying, fighting, cruelty to animals and people, property destruction, and truancy") to genetic causes.[36] The fact that at some point in the recent past character flaws turned into "behavioral disorders" indicates the extent to which individual responsibility and control had already given way to ideas of social and environmental causality. Seeing genes as a determining factor provides another "impersonal" cause. If the individual is in the grip of forces beyond its control, then genes, the environment, abuse, and society constitute a range of mitigating factors that both explain and potentially excuse transgressive behaviors.

This extended chain of behavioral causality plays out and makes visible an underlying metonymy. Behavior becomes the effect of an imaginary chain of causes and influences, from genes to psychology to environment, each cause tangential to the next, as in a mechanical device, but the whole mechanism moving away from less mechanical (and more figurative) processes such as "will" or "responsibility" or even "choice." This suggests that perhaps our increased knowledge of what genes might effect, while scientifically demonstrable, is part of a larger cultural trend that has already shifted causality and responsibility away from the individual to other social, environmental, and now genetic catalysts. It seems only logical that genes might account for impulses explained by social disadvantage, trauma, and mental illness. This is not to criticize the validity of these as causes. Rather, it is to point out how the notions of individual power, responsibility, and the subject itself have changed in the face of molecular biology. The post-DNA subject has perhaps ironically become a composite of forces perceived as "outside" (even though genes are literally inside), a pastiche of virulent fragments, including its genetic legacy and the differential forces that carve its disposition. This subject is characterized by the kind of fragmented and disunified collage that bespeaks postmodernity in the face of responsibility's unified and organized modernity. The individual is envisioned as justifiably having less self-control and ability to exercise choice—and hence bear responsibility.

Just as seeing behavior as an effect of environmental causes seemed to reduce individual responsibility, so the discovery of a hypothesized genetic basis for bad behaviors transfers responsibility for destructive individual malaise away from the social and environmental factors previously thought to affect behavior. The idea that criminals might be a product of various kinds of social and economic inequality fixed part of the blame for criminality on the social and economic inequities of culture. Relocating the cause of bad behaviors to the genes also rather conveniently tends to exonerate culture by naturalizing class and economic conditions as logical extensions of genetically based behaviors. Not only does this obviate the need for governmental intervention, it may be particularly necessary in an age when our most destructive criminals are upstanding advantaged CEOs and white-collar executives. Their genes made them do it. The selfish gene.

This problem of individual control has clearly become an anxiety. The flip side is the increasing prevalence of self-help therapies and fitness regimes that offer an array of self-improvement strategies and a mustering of will to the eroding individual. Obesity, less the subject of speculation about genetic causality than we might expect, becomes the object of a mire of self-control strategies from dieting and exercise to herbs, drugs, spas, and magic cellulite wraps. But recently, obesity has become the site of personal control. Fat becomes the project for television talk shows such as *Dr. Phil* and *Oprah* where weight control is a personal and very public challenge. Fat is overcome by good behavior (which suggests of course that being fat is the result of bad behavior). In one way, individual attempts to control weight make sense. Even if there were a "fat" gene, finding it would have little immediate effect on peoples' girths. But the excessive emphasis on personal will, self-control, and responsible behavior works as an antidote to the loss of control experienced not specifically from genes but from the entire panorama of cultural factors that make it easy to be fat in America. Displacing causality, thus, is less about the possibility of an underlying genetic cause for conditions, as the idea of a causal gene becomes the mechanism that absorbs displaced causality from elsewhere, particularly from the realm of the social. The effect is to exonerate social conditions such as those extensively outlined by Eric Schlosser in *Fast Food Nation* in favor of some mixture of genes and individual will.[37]

The ironic appeal to self-control comes at a time when the subject, who, in a Cartesian system knew itself, now is known by others without its knowledge. Credit reports, shopping patterns, expenditures, phone calls, traffic citations, use of toll roads are all available for tracking and sale. The subject is the effect of information instead of information's controller, transected by messages flowing in all directions, responding on one side to genetic control and on the other to media stimuli. Healed by self-help therapies that envision a pseudoscientific world over which the individual might have some control, individuals seek to regain power through everything from yoga to bodybuilding and dieting regimes to New Age religions. This is not to say that there is anything intrinsically wrong with any of these, only that in the context of an ideological erosion of individual power, the practice of self-improvement conducted by individuals for their own sakes appears very much a compensatory activity necessary to offset our fading sense of agency. The discourses of self-help often suggest a return to nature (or supernature). Using a highly mythologized combination of spiritual connection and pseudobiological potions and behaviors, New Age technologies promise that individuals can either realize their potential through some mode of focus or that they can hook into something more powerful even than genetics and trick nature into a kind of exponential return on its genetic investment.

Anxiety about the centrality of the individual is also evinced by another stream of thought used to explain the genome. Thinking about genes as a repository of human history—as being both the inscription of human evolution and the all-important survival of species information—also subordinates the individual to larger patterns. If the individual is only a single droplet in a genetic pool, then the individual becomes merely a vector for the survival and passage of genetic information. As Dawkins characterizes it:

> Individuals are not stable things, they are fleeting. Chromosomes too are shuffled into oblivion, like hands of cards soon after they are dealt. But the cards themselves survive the shuffling. The cards are the genes [yet another code-script]. The genes are not destroyed by crossing over, they merely change partners and march on. Of course they march on. That is their business. They are the replicators and we are their survival machines. When we have served our purpose we are cast aside. But genes are denizens of geological time: genes are forever.[38]

Dawkins suggests that genes are the reason for human existence and that, even as vectors, humans are not very individualized. Since genes are by their nature not very individual, the same genes or combinations exist and have existed in large numbers of people and species through history. While people have their own idiosyncratic genetic combinations, the genes themselves and their variations exist throughout the biosphere. This means that one's own transient genotype is not in the long run meaningfully unique, nor is one's genetic legacy. If this is the case, parenting becomes less the transmission of an individual line and more a perpetuation of the larger gene pool. If we believe for a prideful and narcissistic moment that we continue in our offspring, we are fooling ourselves. Or perhaps there is a gene for the hubris of parenting, which encourages us to see offspring as chips off the old block when in fact we are all chips from a much larger block. Parental narcissism may be merely a carrot inciting vectors to proliferate other genetic vectors. We may get our genes from our parents, but our inheritance is neither distinctive (except in the case of extreme mutations, which aren't usually good anyway) nor a unique feature of our own family lines. The family is an illusion of an organized and specific genetic line (which is constantly infiltrated with alien genes). What does it mean, after all, that one may have inherited recessive blue eyes from two grandparents, especially when such traits are not necessarily linked to any others (such as personality or stature) or unique in relation to all of the other blue eyes in the world. Phenotypic inheritance provides the illusion of legacy, continuity, and belonging. We want to believe that in all of that genetic complexity, we are unique—and obligingly statistically, we are. But only statistically.

But by the same token, all individuals don't need to reproduce to have their genes passed on. The same genes pass through others, and some genes perhaps should not be proliferated at all. The careful tracking of cystic fibrosis and Tay-Sachs carriers by the Committee for the Prevention of Jewish Genetic Diseases has almost eliminated these genetic illnesses in the Jewish population in the United States.[39] While critics see this as a dangerous eugenics, it could also be understood as a refusal on the part of vectors to proliferate damaging DNA. In the context of humans as gene carriers, we might even view the refusal of some individuals to reproduce—especially if they suspect they carry defective genes—as a heroic stand against perpetuating certain undesirable and certainly selfish genes (even though, as stated earlier, such genes are hardly individual). It is still part of a cultural bias toward reproduction that we regard attempts to limit the passage of bad genes as overreaching eugenics (who is to say what a bad gene is, after all?), and applaud and support the untrammeled passing of any gene whatsoever as a necessary and individually immortalizing cast of the cards back into the genetic deck.

But who or what benefits from this reproductive gloss with all of its myriad associations—individuality, immortality, uniqueness? Humanity, which preserves its pool of genetic information, good and bad? Who knows, conditions we cannot foresee may bring to the fore previously bad genes as an advantage. Mutations, for example, have sometimes provided an advantage, as in the example of the delta 32 gene in fighting the plague.[40] Or we might understand our unquestioned enthusiasm for the reproductive not so much as an instinct for enriching the gene pool but as simply doing what is ultimately better for the genes. Separating genes from bodies in this context is conceptually difficult precisely because we see chromosomal matching and reproduction as the same process. Genetic variety is like a dating service with unlimited choices; it is a resource wherein Nature (or God) knows best. In other contexts, as discussed below, separating genes from bodies is a desirable—even necessary—gesture that requires that we untangle the actions of genes from the insistent scene of reproduction.

But we still look to genes as markers of identity expressed in the physical traits we share with other members of our immediate family despite the fact that genetic research widens its view of the biosphere. We cling to the meaninglessness of statistically unique combinations of our own DNA as a way to underwrite our singular and meaningful existence. Both the Human Genome Project and DNA fingerprinting become ways we can specify the humanity and individuality of our genes in the face of discoveries that DNA is neither too species specific nor very individually distinctive. When the subject becomes an effect, it is comforting to know that we still have our own distinguishing genotypes. Hence the insistence on identity through DNA typing, which continually reaffirms the uniqueness of our genetic combinations and ourselves in a universe that increasingly undercuts difference in favor of variety. And the urgent need to identify, name, and patent genes themselves.

Property

There seem, thus, to be two major, related defensive or compensatory motives for employing textual metaphors in relation to genes: (1) to mask an anxiety about a transition in the dominant logic of signification from metaphor and the law of the father's name to metonymy and the mechanics of digitality; and (2) to assuage an anxiety about individuality. Patenting genes and DNA is another compensatory activity enabled conceptually by genes' textual gloss and further motivated by the promise of excessive profit (another way to compensate the individual). Turning DNA technologies into property represents a complex renegotiation of notions of nature and technology as well as the extension of concepts of commodity and property. The whole possibility of understanding naturally occurring genes as property goes back to concepts imported by textual metaphors.[41]

In general, any metaphor expands the character of the phenomenon—such as genes—that is the subject of comparison. Comparing DNA to language or a text, for example, endows DNA with the capacities associated with language. In thinking of DNA as a language, we imagine that DNA is manipulable, that we can "read" and write with it. It inscribes a "history"; it describes "life." At the same time, language's logic of substitution—word for thing or idea—itself substitutes for the mechanistic operations of organic chemistry. Metaphor, the process of language, masks metonymy, the mechanical process of chemistry. The metaphor of language thus gives us a sense of power over the determinations of biochemistry, a sense that we can read, rewrite, and control a DNA text. If DNA is a text that we can read and if it is something we can transcribe, then it is also something we can copy, selectively manufacture, and alter.

For example, in 1992 a group of researchers from Wales and MIT filed a patent that claimed a property right in, among other things,

> 1. A nucleotide sequence selected from the group consisting of:
> (a) isolated DNA obtained from human chromosome 19, wherein said DNA consists of the myotonic dystrophy gene including a variable number of repeats of the three-base unit CTG or the substantially complementary strand of said DNA, wherein the number of repeats is greater than about 50 in individuals affected by myotonic dystrophy,
> (b) isolated RNA transcribed from the DNA of (a); and
> (c) a fragment of the nucleotide sequences of (a) or (b) which comprises a region of DNA containing the CTG variable repeat region of the myotonic dystrophy gene, wherein said fragment specifically hybridizes to the CTG variable repeat region of the nucleotide sequences of (a) or (b).[42]

On the surface it would seem that this group of researchers is claiming a right in a portion of human chromosome 19, specifically a portion where mistaken transcriptions have led to the pathological proliferation of the three-base unit CTG (cytosine, thymine, guanine) resulting in a condition called myotonic dystrophy (DM) in which a person's muscles atrophy to the point of complete disability. Indeed, claim number 10 on the same patent reiterates this claim, staking out "an isolated human DNA sequence obtained from human chromosome 19, wherein said DNA consists of the myotonic dystrophy gene comprising the CTG triplet repeat region of the myotonic dystrophy gene, or the substantially complementary strand of said DNA, wherein the CTG triplet is repeated about 50 times or more."

How can researchers patent a portion of DNA that occurs naturally? What would enable us even to imagine such a possibility? Isn't such a claim analogous to astronomers patenting stars they discover, or botanists patenting plants they find in the Amazon basin? Naming is one thing (and also a mode of attaching some form of proprietary), but owning? As William Haseltine, the president of Human Genome Science, suggests, "Trying to patent a human gene is like trying to patent a tree. You can patent a table that you build from a tree, but you cannot patent the tree itself."[43] U.S. patent law provides the following: "Whoever invents or discovers any new and useful process, machine, manufacture, or composition of matter, or any new and useful improvement thereof, may obtain a patent therefor, subject to the conditions and requirements of this title." This provides patent holders with "the right to prevent others from making, using, selling, or importing a given invention."[44] In the context of biotechnology, patentable inventions belong generally to the category of utility patents or plant patents. Utility patents apply to new processes, machines, modes of manufacture, or matter composition or useful improvements to any of these. To be patentable, utility patents must be nonobvious, new, and useful. That is, the invention cannot already exist, should be new to the experience of the average person, and must be useful. One cannot patent a theory or an idea.

So how is it that the above researchers could patent a sequence from a gene and why would they want to? Does this mean that if I had DM, I would have to pay them for the use of my nucleotide sequence? What it means is that if I wanted to find a cure for DM, I would have to pay everyone who owned a patent on some applicable bit of genetic information for the use of the information and processes they discovered. How can this be if this nucleotide sequence is not new—is the tree instead of the table? Ostensibly, gene patents apply mostly to genetic tinkering—to processes by which genes can be located, altered, made to work in a different organism, or used in the manufacture of pharmaceuticals or other therapies.[45]

But sometimes, as in the example above, patents seem to apply to preexisting nucleotide sequences.[46] On the one hand, gene sequences, unless they are contrived by biotechnologists, aren't new and therefore shouldn't be subject to patent. But the isolation and transcription of genes seem to be an "original work of authorship fixed in a tangible medium," as patent law requires. What is new, then, about gene sequences is their transcription, the discovery and inscription of the "text" that is regarded as having been authored by the scientists who discerned its order. What is new is what is regarded as genetic "information," which exists as such only because it has been read and translated (note the metaphors necessary even to describe this process).

Reading and transcribing genes usually requires that scientists devise a way to discern the operative, or gene portion, of a DNA sequence by using a marker or some other means of breaking DNA chains into meaningful units. One widely used method is a cDNA (the prefix "c" reminiscent of the copyright symbol) or DNA marker sequences. cDNAs are condensed versions of genes, or "complementary DNAs" that have been derived by trapping messenger RNA active in cells, exposing such RNA to reverse transcriptase (enzymes that chemically return RNA to the DNA form from which it must have come, i.e., if the RNA reads CGUA, the DNA source would have read GCAT). Since RNA exists in nature, what is patented is the DNA work product that resulted from the reversing process. It has been transcribed. In the later years of human genome mapping, this sequencing process was undertaken automatically by computerized machines, and the results were submitted for patent if the sequence was "transcribed" by one of the private corporations such as Celera cooperating in the project. Genes transcribed by publicly funded agencies remain in the public sector.

How any gene could be "owned" by a private corporation and how it could be that some genes are patented and others not is a messy result that comes from shortsighted decisions made in the 1980s about the relation between scientific research, knowledge, and the desire to profit from work underwritten by taxpayers. Two instances of governmental action enabled the acquisition of information that was previously not ownable. The first was the decision in the mid-1980s to permit—in fact, to encourage—scientists to profit privately from publicly funded projects. The availability of property rights in the fruits of scientific discovery emerged as the result of an unrelated zeal in reducing government debt. In 1985 the Gramm-Rudman-Hollings deficit reduction act reduced funding to the NIH and other federally supported research institutions. To cover their shortfalls and keep their labs running, the government suggested that researchers avail themselves of the Technology Transfer Act of 1986, which encouraged government scientists "to collaborate with private industry to speed publicly funded research to the public."[47] Arrangements with businesses were known as "cooperative research and development agreements (CRADAs)" and provided royalties of up to $100,000 per year to contributing scientists. Ostensibly, the act was designed to encourage government scientists to make their discoveries available in a more timely manner, but business interests in exclusivity would work against the free circulation of knowledge. Clearly, then, CRADAs were designed to enable corporate profit at public expense. As Jeff Lyon and Peter Gorner comment, "Government scientists were not only allowed to work out commercial patent rights and exclusive licensing rights to the commercialization of discoveries that the public had paid for; they were supposed to do it."[48]

This all happened pretty much without the general public being aware that their genome was on the block and that information that would have—in a more enlightened time—been public knowledge to be used freely for research and the cure of disease has now become the proprietary interest of corporations who will profit from medical research. If we want to agree that profit is a noble motive without which scientific advances would not be made, then we all also agree to pay much more in the future than we would if we simply paid more tax in the present to retain the public right to scientific information.

But unless the information gleaned through biological research could be patented, there was no use for the private sector to become involved. As E. Richard Gold comments, "Our patent system was designed to promote research through the granting of property rights (to prevent a tragedy of the commons in which no one would invest in research without having private property rights)."[49] What worries commentators is not the basic problem of how a code shared by everyone can become the property right of only a few, but whether the policy of granting patents will ultimately have a deleterious effect on scientific research. As Gold further observes: "Imagine now that different people have the exclusive right to use a particular series of these codes (such as the series that defines a gene). This means that researchers wanting to use, copy, or study (at least in a commercial context) these series of codes can only do so after gaining (usually for a fee) approval from the rights holder. Anyone wishing to study the genetic basis for digesting sugar will thus have to buy rights from a very large number of people—so large a number, in fact, that he or she will simply give up and study something else."[50]

With therapies so potentially profitable, the biotech companies, the government, scientists, and even potential patients saw the private sector as the best way to shepherd lifesaving methods into practice. And the only way to stimulate such development, at least in a capitalist way of thinking, was to permit private corporations and individuals to make money from an exclusive right in a method, a technique, or information derived from scientific research. The "philanthropic" virtues of venture capital were aided historically by a shift in the relation between public funding and business interests.

Hence the second government action was to change the patent law to include specifically the products of biological research. The change was not driven solely by the obvious profit motive but was enabled because for some reason DNA processes were already conceived of as property. Although some DNA sequences may indeed have warranted a utility patent under the old law, both the processes and the actual genes themselves could already be conceived of as patentable because genes have been consistently analogized to texts, and texts are the subject of another slightly different form of property right: copyright. Copyright law reads: "Copyright protection subsists, in accordance with this title, in original works of authorship fixed in any tangible medium of expression, now known or later developed, from which they can be perceived, reproduced, or otherwise communicated, either directly or with the aid of a machine or device."[51]

Patents for gene sequences seem to exist somewhere between the former patent law and copyright. The metaphorically textual gene invites the concept of copyright, while the literally chemical nature of genetic material requires a patent. Without the textual analogy, however, patenting any portion of naturally occurring DNA or its functions would be like patenting black holes and the calculations by which their presence was discovered. The metaphor of the text used for DNA genes permitted the conception of genes as property in the first place. The law simply had to be made to comply with the conceptualization of the gene already in place. If (to argue a negative) somehow the alternative analogy of the template had taken hold in the 1950s, if DNA genes were imagined as topographies or shapes instead of words, our predisposition or unquestioned willingness to transform them into property would have been much more difficult, since (at least for the present) we don't conceive of the shape of land as property.

Patenting genes, however, has not occurred without some soul-searching. "Who owns genes and other pieces of DNA?" asks the Human Genome Project (HGP) Web site in its section devoted to ethical, legal, and social issues (ELSI).[52] That we can even ask a question like this demonstrates a shift not only in our understandings of the body but also in the ways we conceive of property and the possibilities of ownership. If the body is an expression of a genetic text, and if reading and interpreting that text are the rightful property of the scientists who accomplish such reading, then it would seem to follow that these author-scientists would have a property right (a patent) in anything their discovered genes expressed. This is certainly the case for plant patents, cell lines (such as HeLa cells), and certainly specially bred laboratory animals.[53] In humans, however, there is a fine division between the code and the body. The code can be broken up and parsed out to its many readers (e.g., the many different biotech companies and governmental agencies that were engaged in the HGP). The code takes on an existence of its own, detached from the tissue to which it is linked and conveniently so. If the genetic code is a text that can be parsed and distributed among many reader-authors, then its various parts and the methods by which they are read can become the intellectual property of whoever discerns them without having to address the stickier issues of the code's expression in human beings. In fact, the genome deciphered by the HGP is already a theoretical text that represents no individual genome (since there isn't a single human genome). It is instead very much like the imagined ur- or source text deduced from the study of multiple versions of the same story.

Patenting substitutes the right of ownership in the place of the old symbolic law of paternity, which could only really hold sway as long as the father could only be symbolic (i.e., we were never sure who he was). In this way patents perpetuate an older order in the process of its demise. If DNA literalizes the father, then patenting DNA technologies reestablishes a metaphoric paternity in the guise of governmental, university, or corporate ownership of the exclusive right to plunder snippets of DNA information and technologies. DNA patents, which, thus, seem to be only a temporary spur toward the accumulation of information, illustrate the stubborn resilience of paternalistic fantasies in which anxieties about a failure of control or connection are allayed by nominalistic spraying—by appending a name by law to signal dominion. If DNA were to operate freely and without the delusion of proprietary oversight (which, in fact, it does), then what does that do to any human conception of mastery, self-determination, or mystical relation to a deity or superior will? Patents, then, are both potentially lucrative franchises and a way to maintain a fantasy of human centrality and authority (and, quite specifically, the superiority of capitalist organizations). While creating a property relation by appending a text is a traditional way to produce a property right (think of the deed to property), the notion of what can be property has shifted in relation to genetic discoveries and the contemporaneous scramble to copyright information systems.

With the proliferation of digital processors and software, we have come to understand that the "bit" and its organization can be property. Computer and software designers patent and copyright their ways of processing and arranging informational bits, which computer users purchase as computer systems and software platforms. Somehow in all this, the "bit," the little piece of information, which really consists of knowing whether a minuscule switch should be on or off, has been conflated or collapsed into the information it contains. Thus computerized banks of information—names and phone numbers (e.g., your name and phone number)—can be sold as information to a third party who will then use this information to its own ends. In the same way that bits of a person's genome might become the proprietary information of a corporation, so even an individual's personal information—phone number, credit history—becomes the property of those who collect the information. The cost down the line is the individual's, whose time and privacy are buffeted by unsolicited sales pitches or identity theft.

What this extension of the concept of property represents is simultaneously a shift in the conception of what is ownable and a shift in the understandings of what might constitute nature and the personal. While specific bits of "nature" have historically been ownable (i.e., land, *ferae naturae,* trees), natural systems such as species, weather, the cosmos, and physical laws that demonstrate uncontrollable vicissitudes or scale have not been. When, however, biological processes are traced to the level of the micro where there are, for all intents and purposes, orderly and manipulable bits rather than mysterious manifestations of whimsical power, nature becomes technology and microbiology becomes a matter of text and transcription. Like texts and information systems, microbits become ownable, shifting the notion of property from an exclusively macro- to a macro- and microterrain that extends the concept of the tangible from land and gross objects to information and chemicals whose "bit" nature has metaphorically solidified and objectified them. It is difficult to discern whether digitality or genetics or both have pushed the intangible and micro into the realm of the tangible and perceivable; both have arisen simultaneously. However, the digital provides the model of the ownable informational bit that, when transposed onto the textual metaphors of genetics, transforms genes into ownable bits. Of course, this extension of the ideas about what is ownable is also motivated by profit, but by itself profit doesn't shift very conservative conceptual systems such as property.[54]

The New Part

During a 2002 program documenting the importance of DNA genes, a *Nova* interviewer asked Eric Lander if the analogies of the "blueprint, the manual, or the code" aptly described DNA. Lander, who had previously declared the genome "the world's greatest history book," responded: "Oh, gosh, that's a highfaluting metaphor. DNA is a parts list with a lot of parts." Eight months after the June 2000 announcement of practical completion of the human gene sequence, both Lander and Craig Venter, scientist-turned-CEO of the private-sector company Celera, had already revised their claims, or at least the metaphors they would employ to describe the genome. Switching from analogies that implied that the genome somehow provided a plan or instructions, the two scientists suggested the parts list as a way to avoid conveying the idea that they had any idea how the parts went together or how they made the mechanism (the human body) work. "In fact," Venter commented, "what has been said about the human genome, that it is the blueprint for humans, it's not true. We don't think blueprint is the right metaphor."[55] In the *Nova* interview, Lander elaborated his parts list analogy, by comparing the genome to the parts list for a Boeing 707, commenting that with a parts list we wouldn't know how to put the plane together nor would we understand how it flies, but we would be "crazy" not to start with the parts list in the first place.

What motivated this shift from such (at least by 2002) time-honored, grandiose textual metaphors as the "book of life" to such pragmatic scraps as a "parts list"? The change could be an example of simple humility in the face of the extraordinary achievement recently accomplished, or the humble admission that the human genome was never really what proponents claimed it would be. Or perhaps the appearance of the parts list metaphor is a way to prepare for a second stage of research that also needs to be funded but whose support cannot be elicited by using the same terms employed the first time. Or maybe the HGP actually made scientists change their minds about what was to be gained and how it would function. These possibilities, cynical or not, really ask what the purpose of metaphors might be in this context. Are metaphors merely a way to market large-scale scientific projects, or are they attempts to educate a public not equipped with a large scientific vocabulary? Are they slippery ways to suggest that the projects at hand promise more than they can deliver? Or are they merely the enthusiastic wishes of scientists who can see the real value of having such information at hand? Or all of these?

It is, as I have shown (since it has taken two chapters to even get to this point), not a simple matter to investigate the sources and implications of the metaphors employed to characterize complex phenomena. The fact that science spokesmen shift the metaphors at a particular point may provide some insight both into the way such textual metaphors have operated and into why metaphors were necessary in the first place.

Analysis of the data from the HGP in the months following the announcement of its completion paints a very different picture from what in retrospect is the hyped-up rhetoric or even hopes that fueled the funding of the genome's listing. Textual metaphors such as the book, blueprint, recipe, and map abound in material written and proliferated from the discovery of DNA's structure in 1953 to the completion of the first draft of the HGP in 2000. Eighteen months after the completion of the genome was announced, the U.S. Department of Energy, which oversaw the HGP, published "Genomics and Its Impact on Medicine and Society." This 2002 primer represents a moment in which the character of the metaphors used to describe DNA genes and the HGP visibly change. "Although clearly not a Holy Grail or Rosetta Stone for deciphering all of biology—two metaphors commonly used to describe the coveted prize—the sequence is a magnificent and unprecedented resource that will serve as a basis for research and discovery throughout this century and beyond. It will have diverse practical applications and a profound impact upon how we view ourselves and our place in the tapestry of life."[56] Openly switching from two rather grandiose metaphors, one of which suggests a master decoding mechanism, the U.S. government abandons metaphors that suggest that the genome is The Answer in favor of utilitarian promises about what scientists might do with the HGP. It provides the "basis for research" with "diverse practical applications."

Holy Grail to practical applications? Something happened. Or perhaps something had already been happening from the time that Crick first announced in the Eagle Pub in Cambridge that he and his colleagues had found the "secret of life." This transition from metaphors that suggest that DNA is master text to prosaically claiming the HGP a basis for further work, from the idea of the HGP as a single massive opus with everything to calling it a parts list does signal a shift in the logic through which the public is being asked to believe in and understand DNA genes. It clearly happens at this point in history—after the HGP is completed and at the beginning of the next stages of work—to elicit support for the kinds of process and product one can reasonably expect from gene research. But it also represents something else on a more subtle, but not less important scale: a shift in the very logic through which we are now able to understand phenomena. If textual metaphors such as the book, the alphabet, or the code represent the ambition to wield structure and meaning, then the parts list and the breakdown into diverse applications signal the beginnings of a systems understanding of phenomena on a popular level. This shift is anticipated and prepared for by the past fifty years' introduction of digital computers and software logics, but it is only at the point when it becomes completely clear that the genome itself doesn't cure anything that science finally appeals more baldly to these logics. Somehow molecular biology went from Indiana Jones to Bob Vila, and the question isn't so much why it became Bob Vila, but why it was Indiana Jones in the first place.

So we arrive at the parts list, the figure that combines the analogy of the text with the model of the digital bit. In the *Nova* interview, Lander was careful to specify and employ this analogy of the parts list as a humble corrective to the previous few years' textual hyperbole. But what is different between a parts list and a book/alphabet/map/recipe/code/instruction? What the parts list analogy removes is any sense that the HGP provides any order, instruction, or meaning. The sequences of DNA nucleotides, previously vaunted as containing the key to life, are reduced to a collection of elements or bits. Lander's metaphor is still textual in that it is a list, but it has become digital, not as the software analogy but as a bank of facts or data for which the program has yet to be discerned.

The parts list is a hybrid metaphor, one that reveals how work with DNA is indeed in the throes of a transition from a worldview that understands meaning in terms of symbols and stories to one where meaning exists in the ways elements fit together in complex systems. The parts list straddles the two ways of thinking, exposing the shift in the very ways we conceive of phenomena and meaning.

Postscript

Both property and gender seem to reiterate, then, some compensatory extension of metaphor, playing the conventional against systems of variety and dynamism that no longer fit. This does not mean, however, that this metaphoric imperative will prevail forever. Already it is undercut by the very systems—information, chaos, string theory—over which it might try to establish conceptual hegemony. Evelyn Fox Keller, for example, points to how the metaphor of information has enticed genetic research away from the simple cause-effect structures that dominated scientific epistemologies. Because the analogy of information ultimately escapes the textual, it may represent a metaphor that veers away from the metaphoric logic and proprietary possibilities offered by textual metaphors. But it does this at the same time that notions of property are extended to include information itself. The shift of logics is very slow. It may also be the case that genes are the vector by which the transformation from the realm of metaphor into the realm of metonymy is accomplished while keeping in place redefined but still operative notions of paternity and law. The parts list is the quintessential figure of that transition.

Whereas previously in patriarchal culture, children were their father's text in the sense that they were socially identified by means of his name, with DNA and genes both parents become texts and texts become parents. The social functions fulfilled by familial alliances are being supplanted by a sense of the truer script offered by DNA. While "blood" has traditionally been a motivation for alliance, DNA both underwrites and exceeds those alliances insofar as DNA information becomes the site of truth about both individuals and their parentage. This may partly account for the unraveling of the necessary fictions around adoption. The emotional and social ties developed through nurture compete against the "fact" offered by DNA, both in terms of identity and in explanations of behavior. Studies of separated identical twins, for example, have shown that despite environment, twins will act in uncannily similar ways. What prevails is the text, which supplants the parent in favor of truth, on the one hand, and property, on the other. If the text outweighs the parent in symbolic value, how long will the notion of parent last? When do father and mother become a conglomeration of corporate entities with interests in the myriad strands of our genes? At what point will the forms of organization and responsibility whose ideal still governs Western culture be replaced by either the more cooperative and free-flowing dynamic represented by our understandings of information or the repressive competitiveness of the Chicago gangster corporate oligarchy?

CHAPTER 4

The Homunculus and Saturating Tales

Like successful Chicago gangsters our genes have survived, in some cases for millions of years, in a highly competitive world.
—RICHARD DAWKINS, *THE SELFISH GENE*

To turn briefly anthropomorphic, the father's genes do not trust the mother's genes to make a sufficiently invasive placenta; so they do the job themselves.
—MATTHEW RIDLEY, *GENOME*

Genetic Fictions

The popular science writer Matthew Ridley describes the evolution of DNA through the following amalgam of nature and breakfast table:

> The word [DNA] discovered how to rearrange chemicals so as to capture little eddies in the stream of entropy and make them live. . . . The word eventually blossomed and became sufficiently ingenious to build a porridgy contraption called a human brain that could discover and be aware of the word itself.[1]

Ridley's evocations of words, streams, blossoming plants, and gloppy machines provide an imaginative if strangely botanical illustration of the imaginary epic of DNA evolution. The way Ridley arranges his images into a story of purposeful agency, while a shortcut around nearly indescribable complexity, is a typical example of another, less obvious effect of using analogy. Along with the attributes added by other metaphors, DNA genes take on the character of agents. The hierarchical organizations implied in different ways by reductionism, structuralism, and textual metaphors produce the illusion of an organizing agent—in these three cases, the smallest element—who arranges elements and determines their combinations. Atomic elements, tiny structures, and words become DNA actors who busily produce the world around them, including genes themselves. The inevitable effect of agency is consciousness. The DNA "word" becomes a DNA will that is always already an instance of will—of DNA willing itself to knowing agency, so to speak. Ridley's story of the birth of consciousness is a story that endows DNA with conscious power from the start.

Although this characterization of biochemistry doesn't seem like good science and certainly doesn't accord with any theory of the universe, the shorthand of agency and the familiar outlines of narrative are still a central gambit in public accounts of science. DNA genes not only function like language and read like books, but we also regard them as independent agents with human motives and desires that run separately and often differently from our own. Once we conceive of particles as actors no matter how metaphorically, we set them to work in familiar stories of conquest and production, especially and most pervasively in narratives of reproduction (or the intrinsic heteroreproductive impetus of any narrative). This idea of a story is structured around the productive coming together of opposites to produce something else, or reproduce or die or win or know. Any story patterns around this structure. With DNA genes, already endowed with reproductive missions, the seemingly inherent and inevitable heteroreproductive pattern saturates the world of imaginary operations.

Stories with gene protagonists have the full complement of cause-effect relations by which we structure our visible world. Such narratives imagine that the world of the atom or the molecule accords with the Dr. Seuss story *Horton Hears a Who!* in which every smaller particle constitutes a little ordered inhabited universe in itself. Any narrative we might assign to our particulate actors is never, however, really a new story.[2] What we imagine that these tiny agents do is predetermined both by what we think these little agents have already done (which is what we would have done) and by the way stories usually go. In the first instance, the effect—the human body, for example—defines what we imagine genes do. The end retroactively suggests not only the means but an originary (re)productive intention. We have bodies and we have genes; therefore, genes make bodies and wanted to do so. This represents a cause-effect tautology, a circle that results from projecting the end of the story (the body as the ultimate effects of genetic action) as the beginning—as cause or motivation for genetic action.

The analogy of agency smuggles these cause-effect narratives that transform the micro environments of biochemical systems into soap operatic human communities complete with jealousies, vendettas, and willful nucleic acids with human aspirations. Protagonists, conflict, resolution; actor, investment, product; man and woman, romance, baby are all manifestations of a basic narrative impetus to combine opposites to good effect or reduce conflict to a payoff. The unquestioned "truth" of narrative underwrites most of what we conceive of as natural or proper in culture: capitalism, marriage, war. In other words, narrative as a way to organize meaning also reproduces basic ideologies about life, behavior, motivation, values, proper gender roles, and the necessity for closure in meaning, knowledge, production, and reproduction.

When DNA or genes become the protagonists in a story, they also begin to exemplify complex, culturally situated value systems. Not only are they surreptitious narrative agents (or vice versa in that they become agents because they are narrativized), they are also double agents carrying a load of ideological baggage. Anthropomorphic agency inevitably smuggles a raft of ideologies about gender, race, sexuality, species, capitalism, and imperialism. Like metaphors, narrative and its coincident load of human motivations and (re)productive impetus insidiously substitute one mechanism—purposeful cause and effect—for another far more complex and less familiar one, such as evolution or natural selection.

In the history of concepts of evolution itself, for example, notions of narrative and agency have long supplied the logic of purpose that has made it difficult to disseminate Darwin's notions even before the first descriptions of genes. Darwin's theories of natural selection and survival of the fittest are both purposeless. In natural selection, individuals with traits that help them survive a particular environment have a better chance of passing those traits on to their offspring. In a cold environment, for example, members of a species with more fur are more likely to survive and reproduce than those with less fur. Individual creatures don't decide to have more fur. And genes themselves don't decide they need it. Instead, generations of selection increase the number of creatures who have more fur. It is a matter of percentages: more individuals with characteristics that enable them to thrive in an environment will have more children, and their traits will gradually become dominant.[3]

Ideas of agency attributed to anthropomorphized genes substitute will or desire for probability. In narratives of agency, Darwin's finches would will themselves longer beaks or camouflage plumage so they could survive, instead of working as statistical populations. Recasting evolution in terms of genetic agency with human motivation produces two rather unusual but significant tendencies. First, ideologies inherent to narrative's structure, such as sexism and capitalism, become the conditions of genetic organization and human evolution. Second, we come to the end of evolution as we know it, an end that must be quickly recaptured within familiar stories of survival and capitalist success.

Anthropomorphic strategies for representing genes, then, substitute human motivations and psychology for more scientific ways to understand how genes work. They turn genetic systems into familiar narratives and, at the same time, turn genetic narratives into human possibility. These narratives provide a defense against the insignificance of the human individual and the systemic and nonlinear logic of genetic operation. Finally, genetic narratives enable a reconception of evolution at the moment the possibilities of evolution change.

Love Stories

"To turn briefly anthropomorphic, the father's genes do not trust the mother's genes to make a sufficiently invasive placenta; so they do the job themselves."[4] Here come the gender ideologies that are such an intrinsic part of the ways narrative itself works.[5] Even if Ridley's little story, carefully hedged with acknowledgments of its awkward anthropomorphism, is merely a strategy to explain the interactions of chromosomes in the body, its assumptions about gender are not as playful as they may seem. Instead, assumptions about gender roles or their essential qualities have already invaded our thinking about how genes work through both anthropomorphization and narrative. Because gender ideologies are deeply ingrained in both language and narrative, it is extremely difficult to expunge them. Even if scientists believe they are working completely outside ideology (a fairly unlikely possibility, as feminist critics of science have shown, since scientists are human and exist within unacknowledged narratives and ideologies), explanations of genetic operations return to overtly domestic scenarios as the comprehensible medium for all kinds of behaviors, distributions of power, and illogical causality (like father doesn't trust mother) that make no sense scientifically but make automatic sense culturally.[6] Add to that our cultural association of genes and reproduction, and such gendered narratives are doubly loaded with a glistening patina of the innocently natural.

Father's genes not trusting mother's genes, as Ridley suggests, is an anthropomorphization that imports the trite misogynist logic of twentieth-century Western gender relations to partly explain why different sources of DNA govern different parts of fetal development. We might regard an analogy like this as innocent, even humorous play (in a culture where jokes about the old "ball and chain" enjoy a certain currency). But because we actually do not know how this genetic distribution happens, offering an analogy of human behavior from the hackneyed reservoirs of gender wars superinscribes sets of attitudes and motives that tend to offer a psychological or cultural account of *why* genes behave the way they do rather than a scientific account of *how*. The subtle shift from how to why bypasses the complexity of genetic action by offering a familiar, safe gender comedy as explanation. We need not fear the complexities of genes. Genes act like we think we do. Mother's genes are incompetent in the eyes of father's genes, and father's genes are know-it-alls.

In Ridley's scenario, the genes from father and mother act as synecdoches (a part that stands for the whole) for father and mother. This suggests that gendered behaviors reside in every particle of a human being (an idea that curiously returns to pangenesis). Fatherness is expressed in and by every bit of the father, and every bit of a father acts like a father. It also suggests that gender traits and dynamics come straight from the genes, whose behavior in miniature essentially defines the behavior of the whole organism. This is also, then, another example of the circular logic of genetic agency in which the genes that are understood to map traits and behavior are endowed with the traits and behaviors they are said to map.

In a more dangerous way, however, such accounts appear to naturalize cultural stereotypes of gender into the truths of science. If we believe that genes produce behavior and if we project ideas of gendered behavior onto genes, then we end up with the "scientific" idea that genes produce gendered behavior, as such behavior is already culturally defined. If our genes act that way, why shouldn't we? Or worse, what if we think genes behave in pink and blue because some scientist has already imported his gender biases into his conception of genetic activity? We would hope this latter is less likely, but in some areas of scientific investigation, there is always a danger that cultural assumptions about such things as gender will be mistaken for truths about gender to be confirmed by discovering the genes for gender traits—playing dress-up, for example, or liking sports and roughhousing. In other words, these ideas, often barely acknowledged, define results from the start, since we tend to find what we are looking for. And even more, what we are looking for is itself defined by the ideologies that have defined us from the time we become conscious beings.

Although popular culture abounds with instances of genes portrayed as a molecular version of suburban domesticity (see, for example, Ridley's two books), Richard Dawkins had earlier argued for a more serious account of genetic agency in *The Selfish Gene*. Dawkins employs the analogy of selfishness to explain how the proliferation and survival of some genes—the connections between genes and Darwinian evolution—are driven by gene survival rather than by the survival of species or individual "lines." In this model, genes are imagined as the actors or agents, and individual organisms are merely transient hosts. The "selfish gene," Dawkins comments, "hitches a ride in the survival machines created by other DNA."[7] Dawkins famously employs a "Chicago gangster" analogy to account for the kinds of traits we expect from entities that have survived tough turf wars. "Like successful Chicago gangsters," he observes, "our genes have survived, in some cases for millions of years, in a highly competitive world."[8]

Dawkins's recourse to anthropomorphism is an instance in which an ill-fitting analogy is employed for rhetorical, dramatic, and explanatory purposes. Dawkins does remind his readers, "At times, gene language gets a bit tedious, and for brevity and vividness we shall lapse into metaphor. But we shall always keep a sceptical eye on our metaphors, to make sure they can be translated back into gene language if necessary."[9] Despite Dawkins's scrupulously skeptical eye, metaphor and analogy always import connotations and suggestions that cannot be recontained. Like the springy prank snake that leaps out of the box, it is difficult to tuck such an excess of meaning back in place. But the problem, even for a thoughtful author like Dawkins, is that the metaphors we "lapse" into are themselves already predefined and overdetermined. The metaphors we are likely to employ are models that we have chosen in part because of the ideas and values they suggest. In other words, we choose metaphors because they match some notion we already have about a phenomenon. The analogy of language for DNA is one of these, importing ideas of authorship, control, and ownership. Anthropomorphism is another, which imports the foibles (actually the purposed ideological binaries) of patriarchally organized, capitalist, imperialist cultures and gendered, raced, classed, most often Western human behaviors.

It is extremely tempting to envision genes as little motivated directors, as vicious, cigar-chomping, Capone figures with ruthless desires to eliminate the competition. This humanization aligns us with our genes; it enables us to understand genes through a kind of identification with genes as entities with similar motivations. Genes, in other words, are us. This anthropomorphizing also provides a site for an empowered displacement of our own psychical investment from the utter helplessness the human might experience in the face of its inexorable genetic determination and the automatic and inhuman processes of genetic replication. If we can identify with genes, then somehow we are not at such cross-purposes. At the same time, Dawkins's idea of the selfish gene challenges, at least temporarily, human hubris around the discovery of DNA. Genes may be us, but they are not necessarily for us. It is then only by accident that the gene's imaginary agenda and our hope coalesce—in providing a possibility of immortality, as genes see to their own perpetuation.

It's a Small World

Importing human motivation into genetic action projects a world of human scale onto the infinitesimal environment of genetic action, which simply becomes another world like ours. This scalar enlargement tends to dominate popular scientific explanations of genes, which generally involve a series of literal enlargements showing cells, then chromosomes, then an artist's or computer rendering of a slowly spinning double helix ladder featuring color-coded nucleotides in primary colors. At the same time, stories about humans appear in which genes themselves are situated as causal agents. Sci-fi horror renditions, often on a cosmic scale, or sci-fi comedies imagine cloning, identity theft, genetic engineering, large-scale mutation, and DNA mixing in everything from Eddie Murphy's *Nutty Professor* to *Species* (1995), *Alien Resurrection* (1997), and *Gattaca* (1997). These two scales of narrative parallel one another so that the nutty professor's desire to become a macho man itself becomes the gene's motivation for producing macho-ness, or alien genes become secret agents that act on a small scale the way the alien monster works on the large. We might expect the fictions of genetic agency to become the agency of genetic fictions in popular culture. But this also happens in more serious works devoted to presenting scientific knowledge to popular audiences.

In Ridley's two books, *The Red Queen* and *Genome*, melodramatic narratives of genetic conflict and betrayal reveal the dangers of too much anthropomorphism and the infectious logic of popular narrative.[10] Both books invent extended stories about genetic heroes to explain the discoveries made by genetic science. Ridley's tales rely on conventional notions of gender and sexual behavior as motives for genetic action. In contrast to Dawkins's notion of the selfish gene that threatens to overturn ideas of human superiority, Ridley's fantasies tend to reinscribe conservative ideas of gender and heterosexuality as the truth of genetic science.

Ridley's first incursion into genetic reporting, *The Red Queen*, works through the relation between nature and nurture in light of genetic discoveries. The title refers to a process known as the "red queen," named after the chess piece in *Alice in Wonderland* who keeps running to get somewhere, but who can't get anywhere because the environment runs with her. The idea is that evolutionary changes keep up with the environment rather than surpass it. Ridley begins by asserting that the book's aim is to seek "a typical human nature" but that "it is impossible to understand human nature without understanding how it evolved, and it is impossible to understand how it evolved without understanding how human sexuality evolved. For the central theme of our evolution has been sexual."[11] For Ridley sexuality equals reproduction, but because a substantial portion of procreative activity has in fact been asexual, his selection of the term *sexuality* imparts a set of behavioral associations—gender—linked to a specific kind of reproduction (sexual), but which are themselves only tangentially related to reproductive processes.

Ridley envisions sexual reproduction as an evolutionary advance, but his argument is ultimately less about the benefits of sexual reproduction and more about positing modern Western notions of gendered behavior as the natural inclinations of genes. For Ridley sexuality refers to the role of reproduction in natural selection where the evolving gene pool consists of genes from organisms that have survived and reproduced. An individual may survive—that is, live an average life span—but unless that individual reproduces, it will not contribute its genes for successful survival to the larger pool of genes that will constitute the species' genome.

Sexual reproduction, according to Ridley, also implies that there are two "human natures: male and female. The basic asymmetry of gender leads inevitably to different natures for the two genders, natures that suit the particular role of each gender."[12] These natures, Ridley claims, are necessary to evolution, but which comes first—gender or evolution's slow accrued adjustment through time? Does gender (read "sex") enable evolution, or are sexes the effects of evolution? Or both? And doesn't Ridley really mean "sex" instead of gender in the first place? What would be necessary for an evolutionary idea of gender: willing participants in genetic commingling or sets of culturally inflected behaviors asserted as the basis for evolutionary change understood as consisting entirely of successful reproductive lines? The problem here should already be fairly clear. To what extent are ideas about the gendered "nature" of sexual reproduction an observed phenomenon of genetic research as opposed to the imported (and very cultural ideas) about the truth of gender? How do we separate the two?

Ridley's example shows how easy it is to assume the universality of gender ideologies under the influence of narrative, which incorporates these gender ideologies as an intrinsic part of what defines a story. As the basic pattern for narrative itself, reproduction already carries a great deal of baggage: ideas about the character and necessary opposition of genders, ideas about reproduction as the central mechanism and end of any kind of meaning and survival. Our ideas about reproduction, thus, come as much if not more from the shape of the story than from any observed phenomena. The structure of narrative requires binary opposites even if organisms can reproduce in other ways, through fission, for example.

Because of the often unwitting cultural and ideological contributions derived from implied narratives, we need to be careful about the assumptions we make about the essential "nature" of what we understand to be reproduction's components such as a very binary notion of gender and genders' antagonism. Such antagonism may produce the necessary tension for the middle of a story, but it constitutes a cultural explanation that glosses (and obscures) other kinds of mechanisms and systems through which organisms distribute their genetic material such as fission, fragmentation, cooperation, and regeneration. When narratives are evoked as explanatory media, cultural assumptions about the "truth" of gender influence the terms of investigation about gender and substitute cultural for scientific explanations. This has the effect of establishing cultural ideologies as scientific truth. Anthropomorphizing genes, thus, invites a host of assumptions, narratives, and ideologies about the essential nature of things as we perceive phenomena culturally.

Ridley's two books are filled with these anthropomorphic tautologies. Although Ridley seems to be a fairly painstaking and careful reporter, it may be that in narrating human "nature" or the complex interactions of the genome, he borrows what seem to be fairly innocent and accepted narratives, which are never, however, very innocent. Or it could also be the case that Ridley is merely repeating a tendency to borrow narrative assumptions that scientists themselves employ. These assumptions are quite powerful and insidious, involving not only presumptions about the character of gender (it is oppositional, complementary, and pervasive) but also assumptions about the character of sexuality itself.

One example is the way some research scientists account for the idea of "intersexual ontogenetic conflict"—the idea that males and females of nonmonogamous species may produce genetic adaptations that benefit either the male or the female to the detriment of the other sex. One example of this is the observation that the human pelvis has gradually widened, an advantage to women, as it makes childbearing easier, but a disadvantage to men in that it slows down their motor efficiency (though that's not much of a disadvantage these days).[13] Another example of this would be the discovery that proteins introduced with fruit fly sperm discourage females from mating with other males, but are also toxic to females.[14] The benefit of this adaptation is to the male (or actually the male's genes), who keeps his genetic monopoly in relation to other males (but notably not the female) within a system of valuations that projects "fitness" as an ideal of genetic survival.

"Fitness" is defined by W. R. Rice and A. K. Chippendale in terms of reproduction, but differently for males and females—"male reproductive success or female fecundity."[15] Although it would seem natural to think that males and females have different roles in reproduction and hence different forms of "fitness," the ways Rice and Chippendale define fitness differentially actually don't make sense except in terms of cultural concepts of gender roles. Male fitness understood in terms of male "reproductive success" suggests the rate at which males are able to fertilize eggs and, through progeny, pass on their genes. Female fitness is described as "fecundity," a word that connotes a potential for being fertilized (fertility) or receptiveness to fertilization or the actual production of a number of offspring. "Fecundity" is an overly broad, ambiguous term. Why is male fitness described in terms of success and female fitness defined in more ambiguous language? Aren't both successful if the female reproduces?

The difference in terminology reflects gender ideologies. If we understand masculinity to be active, as something that must conquer to proliferate, then the term "reproductive success" means that a male has indeed conquered. It implies a conflict both with other males and with females. "Fecundity," however, is a condition, which implies both a more passive readiness and cooperation. Female fitness is understood in terms of how well she can make the conquering male's dreams come true. These, of course, correlate with cultural notions of male activity and female passivity, male ambition and female help. It would make no difference at all to Rice and Chippendale's experiments or results if they envisioned fitness as "reproductive success" for both males and females, as the genes of each are redistributed equally with all progeny. The problem with using the same terminology is that it simply does not play into the conflict scenario the scientists have mapped, a scenario that takes as its model (at least for the purposes of expression) the hackneyed notion of gender wars so pervasive in Western culture.

That the idea of differential benefits to males and females as a "conflict" is cultural is evident in the way that Rice and Chippendale assert different notions of what constitutes "good" in relation to males and females. They understand male fitness as the male's ability to pass on his genes through fertilization in competition with other males. They measure female benefit not in relation to females' ability to pass on their genes but in terms of a statistical increase in female mortality. Now it is true that mortality will prevent the passage of genes, but at what point? After they have reproduced? And if the female fruit flies die, what benefit is that to the males whose genes have also not been proliferated, at least through the females their proteins kill? A gendered narrative of evolutionary survival has become a cover story for a more complicated process in which genes interact according to the principles of molecular biology rather than Western gender mythologies. What if instead of appealing to a binary model of essentialized gender conflict, we understood the problem of toxic proteins introduced with sperm as a part of a system of adaptations with both felicitous and unfelicitous effects? We might conclude (and what scientists finally do conclude) that these "conflicts" balance out by working through a system of checks and balances that prevent the males from becoming too toxic and the females from reproducing with too many different males. Statistical disadvantages on the scale of small populations become statistically unimportant on the scale of larger populations measured over a longer period of time.

Rice and Chippendale also observe the following: "It was common for genomes that produced the best males to produce the worst females, and vice versa."[16] Again, one can claim this only in relation to cultural notions of what constitutes male and female. Even if this notion has been translated into strictly reproductive terms, the terms themselves are already inflected by ideological notions of gender and its necessary conflicts. If the "best" male is the one who fertilizes the most and the "worst" female is she who is the least fecund, in a systemic view, wouldn't they balance out? Seeing genes as genders and genders as conflicting binaries produces this kind of claim. The point Rice and Chippendale make is that genomes with advantages to one gender tend to produce less successful specimens of the other gender—that genetic gender advantages work to the advantage of one gender or another. This is an investigation into the ways genes battle among themselves, both within an organism and among organisms. But why see this as a binary at all? What if these traits operate beyond reproduction? What if "worse" females are valuable in nonreproductive ways?

The anthropomorphization of the gene produces a familiar scenario of a gender war that is then reinforced by "scientific evidence." What the scientists observe may well be true, but the ways it is characterized are all about ideology—an ideology seemingly naturalized by science—but which deploys gender stereotypes as a starting point. These stereotypes, however, are also comforting, since they provide a familiar cultural logic for the "behaviors" of genes, which reinforces older notions of gender that have recently come into question. If we are worried that women are too much like men as a result of successful feminist movements and the reform of laws, then what better way to reassert a more conservative, patriarchal notion of the "proper woman's place" than to claim that male and female genes behave according to older patterns? And if genes behave this way, that behavior must establish some truth for human nature, since we are our genes.

To illustrate how scientific presentations of antagonism easily become cultural and ideological rationalizations for unhappy gender relations, we need only look to Ridley's gloss of Rice and Chippendale's work: "Suddenly it begins to make sense why relations between the human sexes are such a minefield, and why men have such vastly different interpretations of what constitutes sexual harassment from women. Sexual relations are driven not by what is good, in evolutionary terms, for men or for women, but for their chromosomes."[17]

The idea of this conflict of gender interest is an extension of the concept of sexual dimorphism—that males and females of the same species will evolve differently in relation to their reproductive roles—and that this differential development is indeed natural. We all know that adult males and adult females look different (or do we?), and we link the differences to reproduction and domestic gender roles in general, even though the relation between body development and domestic roles doesn't always make sense. (Who said cleaning was easy? Why can some kitchen cabinets be reached only by tall men?) This dimorphism is sometimes confused with another kind of apparent gendering: the imprinting of genes from one parent or another. Genetic scientists have discerned that in some developmental processes, the gene from one parent takes over the process, while the other gene remains silent or unexpressed. Paternal genes, for example, seem to be responsible for making the placenta, a fact that occasions Ridley's interpretation of fatherly distrust cited above. Maternal genes make most of the brain, while paternal cells are expressed more in the muscles. Maternal genes produce the brain's cortex, the seat of the senses and behavior. Paternal genes in the brain are most active in the hypothalamus and seem to be responsible for social adjustment, leading the anthropologist Robert Trivers to observe that the "cortex has the job of cooperating with maternal relatives while the hypothalamus is an egotistical organ."[18]

Dimorphism and imprinting are two different phenomena, but both fall back on social myths of gender to provide a backstory for evolutionary or developmental differences. It is not that differences don't exist or that genes from one parent or another don't govern certain processes and aspects of development, but that a sociocultural notion of gender has become a way to understand those differences and give meaning to the relation between which genes are expressed and what these genes are responsible for. Analyses that see the outcome of the action of a single gene as reaffirming gender stereotypes naturalize the gender stereotypes they employ in their analysis.

In his investigation of the genome in *Genome*, Ridley suggests finally that binary gender as understood in terms of mid-twentieth-century notions of gender roles is genetic (and therefore natural). "Gender, then," Ridley observes, "was invented as a means of resolving the conflict between the cytoplasmic genes of the two parents."[19] But gender is a cultural idea that resolves many other conflicts, particularly conflicts around the idea of gender itself. In fact, gender is only a cultural idea. That Ridley persistently employs the term *gender* instead of *sex* shows the slippage and ideological infection that occurs because the notion of gender operates in ways sex does not. Genders organize a wide range of differences, variations, and possibilities into a comprehensible dyad (especially where binaries are necessary for cultural sense in the first place). Resolving the question of someone's gender solves a host of other conceptual, behavioral, and social difficulties even, or especially, if a person doesn't easily fit into either gender category. In the United States gender is moored and confirmed by biology where gender and sex collapse into one another. The answer to the gender question (which notably is often the first question about a person posed), if not obvious from appearance or behavior, is made by recourse to genital morphology and more recently to genotype. Gender thus serves as a kind of "truth" that untangles and organizes the fact that gender itself is neither so clear-cut nor obvious as we like to think.

In his Solomonic solution to the question of whether gender (not sexed) behaviors are the result of nature or nurture, Ridley suggests that it is clear that gendered behaviors come at least partly from nature. A predisposition to dolls or trucks, he suggests, may be genetic. "Boys and girls have systematically different interests from the very beginning of autonomous behavior. Boys are more competitive, more interested in machines, weapons and deeds. Girls are more interested in people, clothes and words."[20] Such sweeping generalizations about gender reiterate pre-women's movement rhetoric about the truth—even the divine intention—of "proper" gender roles, licensed at this point in time by an appeal to "nature." "Nature" in Ridley's discussion actually means genes, and since the genetic world is imagined as our world in miniature—that is, culture *is* nature—then little genetic actors can be expected to act as we would and to reflect our biases, predilections, and ideologies. An appeal to nature in the twenty-first century revives retrogressive binary understandings of gender both as a rhetorical mode of simplification and as a response to fears that the truths we have long been accustomed to will disappear with an increased understanding of genes and DNA. Although the possibilities of genetic action clearly produce a spectrum of variations, popularizers ignore the spectrum in favor of the conservative binary, especially in the arena of gender, which seems so basic and necessary to our narratives of reproduction.

What happens if we perceive gender without the binary blinders? What about the girls who like trucks and the boys who like clothes? The first instinct is to relocate such individuals as "inverts" in a reproductive scheme—as homosexuals who simply occupy the opposite sides of the same binary. But what if these are really different genders? The science historian Anne Fausto-Sterling suggests five genders in her sustained critique of science's participation in gender ideology. "Labeling someone a man or a woman is a social decision," Fausto-Sterling comments. "We may use scientific knowledge to help us make the decision, but only our beliefs about gender—not science—can define our sex. Furthermore, our beliefs about gender affect what kinds of knowledge scientists produce about sex in the first place."[21]

Noting that scientific truths about sex are a part of the larger cultural struggle, Fausto-Sterling points to the existence of several varieties of genetic and developmental gender. Focusing on the rigid conformist treatment of intersexuals (those who manifest some combination of male and female morphologies and hormones), Fausto-Sterling makes clear how such conditions are pathologized because such individuals are incomprehensible within medicine's notion of the "truth" of binary gender. Treated with surgeries designed to align gonads and genitals on a single side of a gender binary, intersexuals are essentially denied their intersexuality in favor of gender conformity. What is "natural" here? Clearly, intersexuals are a genetic possibility. They occur naturally and without help. Why must they be made to conform to an artificial binary outside the fact that culture cannot comprehend anything beyond two genders?

What if we argue that there have always been more than two genders, especially if, as Ridley suggests, gendered behaviors are also genetic? Not only might we have the five sexes Fausto-Sterling suggests—males, females, herms (true hermaphrodites), merms (male pseudohermaphrodites), and ferms (female pseudohermaphrodites)—but also the various combinations of biological sex and gender behaviors (tomboys, queens, androgynes) aligned in various ways with Fausto-Sterling's five. The point is not to have an endless list of gender permutations, but that sex and gender is a system with multiple possibilities we already understand and incorporate socially. In other words, culture is actually more perceptive than science, especially a science crippled by stubbornly simplistic assumptions about the sex-gender system. As Fausto-Sterling contemplates: "I imagine a future in which our knowledge of the body has led to resistance against medical surveillance, in which medical science has been placed at the service of gender variability, and genders have multiplied beyond currently fathomable limits. . . . Ultimately, perhaps, concepts of masculinity and femininity might overlap so completely as to render the very notion of gender irrelevant."[22]

Part of the reason simplistic binary gender is so constantly reiterated is the insistent character of our narrative of reproduction that can comprehend reproduction only as the conjoinder of opposites. This narrative is quite influential in defining binarized genders as the actors in this drama. Even if we can perceive and acknowledge that multiple, nonbinarized genders exist, we have difficulty fitting them into our reproductive story. Instead, even culturally, we tend to try to squeeze nontraditional individuals and arrangements back into the story—conforming intersexuals and expanding the notion of family to accommodate the potentials of fertility technologies. A resistant aspect of the insistence of this particular reproductive narrative is that it is at the center of our understandings of evolution. For this reason, though we might assail gender, it is extremely difficult to bring our narrative about reproduction into question. Instead, we employ the narrative of reproduction glossed as the narrative of the patriarchal nuclear family as a template to understand the systemic interrelation of genes, complete with gender wars, parental influence understood within a very social gloss (e.g., the father's genes control the development of an "egotistical" part of the brain), and sexuality itself. To operate within a particular understanding of reproduction, complex behaviors, such as gender attributes that can at best be only cultural, must also be made genetic.

That ideas about reproduction and evolution are interwound provides one reason for the ways gender and sexuality are seen increasingly as genetic, while race becomes less so. But the trend toward geneticizing gender and sexuality also serves particular cultural and ideological purposes, which seem, in fact, quite at odds with one another. The genetic naturalization of binary gender accompanies the discovery of the utter complexity and multiplicity of sex and gender. Rediscovering the "truth" of a desired gender simplicity is a defense mechanism against losing the kind of imaginary order provided by social systems premised on binary gender. The patriarchal nuclear family may be waning, but science is reviving it as a truth on many levels.

Fruit Flies

If complex gendered behaviors can be imagined as genetic, then those behaviors become natural and therefore right. Or so the argument might be in relation to the rather complicated inmixture of enlightened humanism, scripture, and science that dominates the American public imagination. In this imagination (worthy of an entire study itself), God's will is executed by the mechanisms discovered by science.[23] God's will is fairly elastic, but usually consists of the patriarchal nuclear family longingly resurrected, with goodness the norm and evil the result of some kind of mistake. God may have created the heavens and the earth, but he accomplished this via the big bang. God created the creatures, but he did so through DNA and evolution or, in more insistently scriptural regions, "intelligent design." God made sinners and saints, but he may also have accomplished that through genetics. And herein lies the dilemma: if aspects of character we understand to be good or bad are the effects of genes, then in what way can sinners reform? Or are genetics merely the mechanism of Calvinist predestination?

Clearly, this is an extremely simplified version of how science has been integrated into belief systems by a population systematically undereducated in science. Issues of ethics and blame circle around the individual and especially around problems of causality. If antisocial behavior is genetic (as asserted in the case of XYY males), then what blame must they bear for their actions? Genetic variations are likened to other conditions such as mental defect, which are seen as preventing individual responsibility. If an individual is genetically predisposed to obesity, then that person's culpable fatness might no longer be seen as the unfortunate effect of uncontrolled gluttony (I use this example because obesity is understood as weakness in American culture). And if one's child is gay, then perhaps that homosexuality is also caused by a gene instead of myriad and mysterious social factors, including family environment and parental behaviors. If genetic, then gayness is not the result of choice (after all, who but the evil and antisocial would choose such a thing?).

Understanding behaviors (and desires) as genetic seems like an answer to many questions. First, it reduces the complex phenomena of homosexualities to a single phenomenon; there are many kinds of homosexuality, not to mention (as most don't) various kinds of lesbians. Second, if homosexuality is genetic, then it is somehow "chemical." Third, it displaces any blame for homosexuality into the impersonal action of DNA, "naturalizing" homosexuality. This naturalization, however, is where ideological problems arise. If homosexuality is "natural," then aren't discriminations against homosexuals unethical and unjust? And aren't family structures and all other social organizations premised on heterosexuality a bit one-sided and themselves against nature? If we regard homosexualities as genetic and therefore natural, should we expect homosexuals to change, or should we consider eliminating them through engineering? Should we discriminate against them (a practice that is still legal throughout most of the United States and gleefully defended by some of the more avid defenders of the faith)? Could we identify homosexuals from their DNA?

If we have been crudely successful in aligning traditional ideologies with genetic discoveries, homosexuality proves to be the sticking point, for with the queer, genetics is damned either way. If homosexualities are not genetic, then we are somehow culpable for producing them, even if we displace that culpability onto the individuals so afflicted. If homosexuality is genetic, then behaviors can't be changed, and God's plan clearly includes queers. But if it is genetic, we might also be able to engineer it away as if it were a genetic disease. But homosexualities do not fit into the paradigm of disease, since, unlike cystic fibrosis or other genetic conditions, homosexualities neither disable nor kill people. The case of homosexualities demonstrates the biases of culture in relation to declarations of genetic "nature," both insofar as complex behaviors can be declared to be genetic and in the dilemma of how to regard the idea of naturally occurring variation.

If homosexuality is genetic, then wouldn't heterosexuality be also? Why aren't we looking as hard for a "straight" gene? The issue of homosexuality occupies a small but overdetermined site in the press politics of genetic research reportage. Whether homosexuality is genetic has a bearing on how we understand the righteous naturalness of heterosexuality and all that follows from it, including the unquestioned mechanisms of our ingrained narrative of reproduction. If evolution requires reproduction as a necessary ground state for species survival, how do species survive if they regularly produce homosexual behaviors? How might it be that homosexualities survive, if we understand homosexuality to be nonreproductive? How would "homosexual" genes be perpetuated? Questions about the genetic basis of homosexuality reveal quite explicitly the anxieties and investments in the close relation between familial and gender ideologies, concepts of reproduction, and science.

Homosexuality provides one point where the ideology of science becomes quite clear (i.e., if some scientists' uncritical adherence to gender ideologies is not clear enough). But the point about gender is that its status as ideology really isn't clear to scientists when they substitute culture for scientific observation, a substitution easily accomplished, since gender is completely ingrained in our language and thought systems. Because of how we understand reproduction to be necessarily the productive conjoinder of opposites, however, the fact that homosexualities are also reproduced suggests that perhaps our heteroreproductive story is not so all-encompassing after all.

Of course, homosexuality could be simply a recessive trait, one whose appearance doesn't really disturb ongoing species production. But why is it that we assume that homosexuals don't reproduce? Have all of those spinster aunts and bachelor uncles somehow immaculately invaded the gene pool anyway? One problem, of course, is how one defines homosexuality: is it behavior, desire, or an orientation imprinted in the brain? The inclination has been to see homosexuality as a behavior, since the notion of behavior is imprecise and versatile enough to absorb a category as complex as homosexuality. Because homosexuality has long been the object of psychoanalytic and sexology studies that occur roughly in the realm of behavior, it had already achieved status as one of the central "deviant" behaviors that is both natural (or normal) and perverse (out of sync with supposed truths of nature).[24] But what behavior is it? There are many vestiges of homosexual behavior, too many variants even to list, starting with the vast differences between gay men and lesbians. There is in fact nothing that prevents homosexual individuals from reproducing, and they do, which suggests not only that homosexuals can produce heterosexual children (which they do) but that the opposite is already true: heterosexual reproduction produces homosexual offspring. The effect of recognizing that homosexuals reproduce suggests that reproduction has never been as patriarchally familial as cultural narratives would have us believe, nor perhaps is the survival of genes so thoroughly the product of a ruthless program of fitness survival. This becomes particularly visible at this point in time when technology openly alters what "fitness" might mean.

Precedent for understanding behavior as genetic came from Seymour Benzer's experiments with fruit flies. Zapping flies with X-rays, Benzer discovered that certain aspects of fly behavior, such as their tendency to go toward light or the timing of their daily rhythms, would alter. One mutant, the by-product of a search for sterile female mutants, was a male mutant with a tendency to chain dance with other males. The graduate assistant who discovered this mutant named it "fruity" and published a short note on its characteristics. Jeff Hall, an associate of Benzer's, read the note and decided to investigate what relation the fruity mutation might have to a courtship gene. Hall, in a gesture that strips science of its neutral pretense, renamed fruity "fruitless."[25]

The drosophila geneticist Hall and his associates worked out what had happened with the fruitless mutation. It turns out that a piece of DNA from "fruitless" "had popped out and then gotten stitched back into its chromosome backward. The X rays had caused an inversion. . . . In effect, fruitless is two mutations close together, one at each end of the inversion. Some of fruitless's behavior maps to the break point at one end of the inversion: the fruitless male's failure to copulate and his habit of chaining with other males. Another piece of fruitless behavior maps to the other end: the male's ability to stimulate courtship in other males."[26]

Displaced into the fruit fly and dressed as science, the genetic basis of a behavior interpreted as homosexual (and hence "fruitless") turns out to be an "inversion," probably not uncoincidentally the same term used by late-nineteenth-century sexologists to understand mechanisms of homosexuality. Havelock Ellis, for example, saw homosexuals as beings whose genders and desires were inverted in relation to their bodies: male homosexuals were women in men's bodies, and lesbians were men in female bodies. Hall's investigation turns sexuality into a kind of DNA grammar, composed of behavioral components such as a "failure to copulate," "chaining with other males," and an "ability to stimulate courtship in other males" that has literally been inserted backward. Genetic science ratifies Ellis's sex-gender logic.

If we can find a simple gene inversion in fruit flies, why not look for such a genetic "mutation" in humans? Although human sexuality is an extremely complex set of behavioral and chemical dynamics that may change throughout an individual's lifetime rather than a single "condition," researchers nonetheless looked for genetic or morphological differences that might distinguish homosexuals from heterosexuals. The first results in 1991 suggested that in identical twins, if one was homosexual, the other was more likely to also be homosexual than was the case with fraternal twins. Simon LeVay of the Salk Institute also discovered that a portion of the hypothalamus was smaller in gay men than in straight men. Presumably the difference in morphology is linked to homosexuality and is produced by genes. In 1992 more brain comparisons produced the discovery that a portion of the fibers connecting the left and right sides of the brain was larger in homosexual men than in heterosexual men. In 1993 and 1995, Dean Hamer of the National Cancer Institute discovered a gene on the X chromosome, Xq28, that seemed prevalent in homosexual men and not in heterosexuals. When another scientist, George Rice, tried to reproduce Hamer's work, he could not find any greater incidence of this gene.[27]

At this point in time, then, the whole question of a genetic basis for homosexuality is up in the air. It may be that there are genetic variations that produce the behavioral phenomena we interpret as homosexuality, but given the complexity of the variety of behaviors and the interrelation of behaviors and gender norms, it is most likely that there isn't a single "gay" gene. As the biologist Ruth Hubbard observes,

> I think we are talking about a psycho social sexual complex of behaviors that get defined in certain ways by certain societies in certain historical moments. Behavior that may be called homosexual at one time, or by one kind of society, will not be called that at another time or by another society. I think we're talking about social characteristics rather than biological characteristics when we talk about homosexuality. And that biology may get into that . . . oh sure we're organisms . . . aren't we? . . . yeah and we have sex organs and yes there's biology in everything we do. So to that extent sure there are gonna be genes in it. But so what?[28]

So what indeed. The issue here is not so much that homosexualities might have a genetic basis, but how our belief in that genetic basis might shake up a cultural system premised on a truth of binary gender within the imperatives of reproduction. The effects of scientific inquiry into the possible genetic bases for homosexuality expose the anxieties and uncertainties that attach to this point. Some gay rights advocates welcome scientific confirmation of a gay gene, since finding such a gene would naturalize homosexuality and relocate it as an analogy to other "natural" differences worthy of legal protection (such as race). Other gay rights activists eschew science, claiming that there is nothing wrong with being gay in the first place and that gayness deserves rights anyway. Religious fundamentalists bent on preserving the patriarchal nature of the nuclear family also want to deny a genetic basis for homosexuality because a genetic cause suggests the impossibility that gay people can "change" into heterosexuals. If the various versions of homosexuality are indeed "natural," what does that do to our conception of the necessary and sacred rightness of patriarchy?

Passing On

One problem with the idea of the gay gene is that it is difficult to know politically what positions to take about biological causality. Which is more desirable? A gay gene or no gay gene? Homosexuality as "hard-wired" and natural, as environmental, or as voluntary? Science and politics clearly intersect at this point, where the stakes in considering a disposition "natural" or voluntary are themselves quite contradictory. If voluntary, for example, then gayness could be changed, and it should be accorded no rights under the model currently employed for racial minorities. But if there is nothing wrong with gayness, then why shouldn't gays and lesbians be accorded the same rights as everyone else? The whole question of the "cause" of homosexualities is bound up in the relation between science, human rights, and ideology. At this point in time, any question about gay gene-ality exists on the margins of a larger set of issues around reproductive technologies and a relation between the "right" to reproduce and the operations of evolution. Up to now, a superficial understanding of evolution would suggest that the fittest survive and pass on their genes for fitness to their survivors, who then reproduce, and so forth. One capacity necessary to this scheme is the ability to reproduce. Reproductive "fitness" would be one element of the survival of the fittest. But what happens in an age when we engineer the reproduction of the reproductively unfit?

In twenty-first-century America, reproduction is cast as an inalienable right. That is, no one is prevented from reproducing, at least not ideally or legally. It would seem that the horrors of eugenics have prompted the protection of a reproductive ideal that has become absolute. Those who cannot conceive domestically but who have the fiscal means may try a battery of fertility techniques from in vitro fertilization and implantation to sperm banks, egg donors, and surrogate mothers. Fertilized eggs can be manipulated to increase their chances of proper development, and fetuses can be checked for potential problems through amniocentesis. Technology enables the reproduction of those who might not otherwise be able to reproduce, as well as the survival of fetuses and infants who formerly would not have survived.

What do reproductive technologies do to our concepts of evolution as the survival of the fittest through reproduction? Those who were previously "unfit" are surviving and reproducing as well, including an increasing number of gay couples. If we understand fitness in relation to the environment in which such fitness is measured, then in fact those who can reproduce with technological help are quite fit. But don't we still think of fitness in relation to some imagined "natural" (i.e., nontechnological) state? In less technological conditions, those who need technological support would not reproduce and some would not survive, but for a variety of reasons beyond the question of fertility.

Although there is some discomfort around fertility technologies (the law, for example, has had to adjust to the reality of the variations from patriarchal assumptions), no one really questions the right of individuals to seek fertility help. There is, however, a continued battle between social understandings of "family," which can expand and work despite a lack of genetic connection among its members, and an insistence on genetic connections as adopted children and the offspring of sperm donors aggressively seek their "biological" parents. These notions of relations both compete with one another and coexist, playing out the tensions between the social and the scientific. But the relation that tends to win out is the "right" to know who one's genetic forebears are. The imaginary of the "bloodline" is bound up in the irresistible narrative of immortality. The denizens of the well-fed environs of industrial society are thus loath to consider how technology might alter evolutionary processes (and perhaps they are right, since we have no idea how temporary our technological boon might be). Rather, they are driven by the fantasy of a gene agent literally perpetuating individuals by investing and expressing itself in another body, producing ideally a line of bodies who share the same genes. The gene works as a synecdoche for the survival of the individual, despite the fact that the inmixing of genes can produce a great deal of difference and that the genes of the nonreproductive can also be passed on and expressed through other family members.

Race Erasure, or The Taxonomic Imagination

Ridley is writing his science fictions at the turn of the century because of popular interest in the Human Genome Project, hoping to make accessible some of the complex discoveries of the past twenty years of genetic research. Popularizing seems always to require some degree of romanticizing, and, of course, the best "romance" is romance. Even DNA gets a love life. But at the same time that we start imagining genetic cotillions, race as a genetically based category is declared to be genetically nonexistent. The question is why, when race is happily announced not to be a genetic truth, does gender become a central truth? This is not to suggest that somehow we need to reify race, but that a very traditional and certain notion of binary gender comes to compensate for the loss of other categories of differences, however cultural they might always have been.

The word *race,* in use since the sixteenth century, has an obscure origin and refers to multiple concepts. It can signify the offspring of a single person, a particular group, a tribe, nation, or people, a large and significant division of living things (i.e., the human race), a species, or a set of beings who share common characteristics. The concepts associated with the term are, superficially at least, all associated with a projected commonality and the reproducibility of common traits. Although we might consider such categories as a species to be defined in terms of what are sets of common characteristics, species are significant only insofar as they are based on one primary difference: the tendency to reproduce only among themselves. The significant difference represented by taxonomies is primarily reproductive, even though reproductive unavailability is often identified on the basis of morphological or structural differences. Differing species of finch have differently shaped beaks. A horse looks different from a donkey. This notion of reproductive exclusivity within species holds sway despite the fact that many species can and do interbreed.

Taxonomies are almost always hierarchical and are produced in relation to a set of assumptions or protocols about what might constitute significant differences. Taxonomies, like almost any classificatory scheme, generate their own truths by virtue of the scheme itself. As a taxonomic concept, race relies on a belief in the constancy of inheritance and the separability of types or classes of organisms through evolutionary change. Classes, often defined by cultural rather than scientific norms and through such processes as morphological analogy and homology, produce the appearance of commonality and thus the truth of affiliation, the perpetuation of associations, and a belief in the inescapability of type. Affiliation presumes shared ancestry, an evolutionary link that suggests that our current taxonomy is both structural and historical, reflecting the evolution of species through time.

The current taxonomy of species devised by the eighteenth-century physician and botanist Carolus Linnaeus (1707–78) is based on plant reproduction. Because plants' reproductive parts seemed to remain rather unchanged through the course of earth's history, Linnaeus used the parts—and a colorful latinate genital vocabulary—as a way to classify plants in general. This taxonomy developed into the hierarchical system we have today where different types are defined by their structural and categorical similarity to other types as well as their location within larger more grossly associational "kingdoms," "phyla," "subphyla," "orders," and "families." The choice of names for this hierarchical line of belonging reflects the more ideological investments that adhere to such a scheme: divine-right governments, Greek administrative tribes, governed organizations, and patriarchal units. The scheme's neatest invention was its adoption of patriarchal naming practices wherein each species would be identified by a "genus" name that located it in the larger taxonomic scheme and a "species" name that identified its singularity (often recalling its discoverer as a kind of second father). The taxonomy of all living things in itself imported ideals and assumptions from human organizational practices—ideals about organization, permanence, law, and truth.[29]

In relation to Linnaean taxonomic order, race is a fictional taxonomy arising from an imaginary version of evolution and the misapprehension (or ideological application) of biologically insignificant differences. Just as Linnaeus's hierarchical tree assumes degrees of relation based on an imagined scheme of evolutionary descent—humans are more related to chimpanzees than to trees because the common human-chimp ancestor is closer in evolutionary history than the common primate-tree ancestor—so rendering race a significant taxonomic category suggests some evolutionary bifurcation. Either differently raced people split from a common ancestor at some point in time, or white humans have descended from more "primitive" black and Asian humans. The differences, which reflect geography rather than such a definitive evolutionary history (or geography through fairly recent history), are premised on vaguely morphological but also curiously behavioral attributes. If Linnaeus can classify plants based on reproductive structures, race is the cultural classification of humans premised on the projection of cultural values interpreted as behavioral characteristics imagined to be as common and reproducible as all other species' characteristics. And in keeping with the reproductive bent of taxonomies, the subsequent intermingling of races, like species, is imagined to result in something like mules.

Eugenicists such as Charles Davenport (1866–1944) believed that such traits as feeblemindedness, sloth, and criminality were inherited, as were traits for strength, courage, and intelligence. Just like his eugenic predecessor Francis Galton, Davenport believed in the possibility of breeding a superior human and tried to develop mathematical formulas for the appearance of certain traits based on elaborate family histories. Davenport's contemporary, Earnest Hooton (1887–1954) was a physical anthropologist who believed that there was a correlation between physical traits and cultural and racial attributes. Both Davenport and Hooton relied on an assumption that some agency—genes—was reliably responsible for reproducing a large range of behaviors and traits associated with bloodlines, physical morphologies, and social conditions. This resulted in the "scientific" ratification of racism—the "objective" reification of a fictional category as an evolutionary and genetic truth.[30]

The recent declaration of the nongenetic nature of race, then, is an overdue correction to what had been a long-standing taxonomic fiction that confused social, cultural, and ideological causes for genetic effects. Not only is race not genetically based, the history of the development of racial "types" is not as clear or direct as previously believed. Beliefs about the distribution of human traits are primarily geographic; that is, they depend on our Darwinian belief in the relation between environment and morphology. The migratory movements of people across geographies, however, produce both environmental and political change that results not in immediate genetic adaptation to new environments but in the blending of traits between the newly intermixed peoples—or not. Failure to blend is an effect of politics rather than nature in the form of antimiscegenation laws, the maintenance of rigid class systems, the doctrinal sequestration of religious groups, or the perpetuation of exploitable economic deprivations. When the ideological imperative for such divisions disappears (or becomes less urgent), the divisions collapse if they hadn't already, biologically at least.

Race, in fact, is only the most recent casualty of genetic science's clash with traditional ideology. Luigi Cavalli-Sforza had previously taken apart assumptions about national and ethnic genetic differences in a series of lectures delivered in the 1980s in France. Published in 2000 as *Genes, Peoples, and Languages,* Cavalli-Sforza's work studies the statistical differences occurring in selected gene markers in relation to language groups.[31] What Cavalli-Sforza shows is not only that migrations of populations can be traced through statistical genetics combined with linguistic analysis (in another but more useful example of the correlations between genes and language) but also that the differences among people are better understood as a continuity rather than as difference. As Cavalli-Sforza rather oddly grants: "The intellectual interest of a rational classification of races clashes with the absurdity of imposing artificial discontinuity on a phenomenon that is very nearly continuous."[32]

Like any scientific inquiry into large, cultural phenomena, Cavalli-Sforza relies on cultural notions of what constitutes a population, even though part of his conclusion is that those cultural notions are not based on fact as defined by the incidence of gene markers. As is true with any scientific interrogation of the "naturalness" of gendered behaviors, inquiries into the naturalness of any cultural category will end up at least partly reifying the very category under investigation. This was true of inquiries into the biological bases of race until Ashley Montagu's 1942 book *Man's Most Dangerous Myth; The Fallacy of Race.*[33] Like the work of most thinkers ahead of their time, Montagu's commonsense argument did not arrest continued investigations into the truth of racial difference, nor did it alter cultural attitudes premised on the symbolic exclusion of those deemed "other." Nonetheless, gradually and finally with the sequencing of the human genome, science again comes to the same conclusion: there is no genetic basis for race.

At the same time, medicine asserts a continued use for racial categories insofar as people of the same groups tend to have similar configurations of certain genes that may predispose them toward certain diseases or define their reactions to certain treatments or medications. This is admittedly a statistical rather than essentialized tendency, especially since what might constitute a "racial" group is difficult to discern and may itself be determined by the same racist criteria genetics studies have dispelled. For instance, according to Byron Spice of the *Pittsburgh Post-Gazette,* "African Americans seem to have a higher frequency of one mutation that reduces the liver's ability to break down certain tricyclic antidepressant drugs, so they are more likely than their white counterparts to suffer side effects. Blacks also have a higher frequency of a mutation that increases the speed at which a newer class of antidepressants, such as Prozac, take effect."[34] But not all African Americans are African in the same way (i.e., African already stands for many different populations and genotypes), nor are African Americans necessarily treated with the same aggressive vigor and enthusiasm in the first place.[35]

Of course, getting rid of race makes a great deal of sense, given the idea that race is a cultural construction with an ideological overload (as is gender) rather than a phenomenon observable in discrete, absolute, and clearly marked differences other than those that adhere to many historically close groups of people. As information about the human genome is disseminated to the larger public, significant differences and similarities, whether true or not, are bruited about: There is no genetic basis for race. Humans share 98 percent of their genes with chimpanzees. There is a homosexual gene. Gender is genetic, and not only genetic, but also central and operational. The two-contributor model on which human reproduction functions may be something we are beginning to understand on a molecular level, but the essential gender attributes somehow also imagined in the processes are cultural and not genetic. A gene from the father, for example, doesn't need to be aggressive, nor the mother's gene passive.

The pattern of certifying gender while we eliminate race is not a question of scientific fact but more an issue of which cultural categories must be preserved to salvage and perpetuate at least a vestige of familiar cultural systems. Race is no longer necessary in a world of diverse variety either to sustain the white man's sense of himself (well, most) or to underwrite labor systems. Diversity (or multiculturalism), the oft-lamented terminology of "political correctness," simply recognizes on a basic cultural level the coexistence of multiple varieties of people and cultures. At the same time, the institutionalization of diversity is at the behest of white endowers whose acknowledgment of the human condition somehow becomes heroic and self-sacrificing. For those who do not accept such ideas as anything other than liberal dreaming, diversity and multiculturalism are the code words for a continued double standard in which liberals are giving away opportunities to the less able. In other words, for those to whom the idea of diversity is a welcome insight, the discoveries of genetics merely reinforce proper social attitudes (even if on some level such reinforcement is a variety of abolitionism). For those whose sense of importance must be reinforced by a comparison with others imagined as different and inferior, declaring race not to be genetic is simply another way to mask the evolutionary supremacy of white Europeans. And they will find scientists who agree, just as fundamentalists who oppose evolution will find (and have found) scientists who will ratify some aspect of creationism (see chapter 5).

It is ultimately a wise move in any case to acknowledge the genetic sameness of nonwhite citizens of Western countries as they continue to become more visibly enfranchised. I say this ironically only as science's proclamation comes (actually is heard) belatedly and at the very moment conservative factions pressure for the end of programs that help disadvantaged citizens. While it may be nice finally to acknowledge that there is no genetic difference, that recognition comes at the very point when no difference serves a fairly conservative political agenda. If there is no genetic difference, then why the "special" programs? Why affirmative action? The confusion between imagined racial differences and social disadvantage has enabled both discriminations ("blacks are poor because they are inferior") and the desire to end programs that redress those discriminations ("blacks don't need help because they are the same").

Nor is race still necessary to abet systems of production and exploitation. In an era when most large corporations operate transnationally, there is an ideological imperative to imagine a one-world idea about populations. This, of course, does not reflect the disparate conditions under which these populations exist—that is the whole point. The moment in history when poor populations begin openly to serve as the locus of labor for capitalist interests that thrive elsewhere is the moment when the treatment of those populations is masked by declarations of their coequal humanity. In other words, any exploitation of nonwhite workers is expiated symbolically through the scientific admission of their human equality.

For some, however, this scientific expansiveness is itself suspicious. From a Web site:

> The American (Euro) scientists who created AIDS believe the anthropology textbooks that indigenous groups of people who inhabit all parts of the global world are Caucasian. The scientists, therefore, felt safe to develop a disease that would wipe out all people with Negroid Ancestry. Little did the scientists know that only 8 percent of the global world is Caucasian—WHITE! Many of the indigenous people (now classified as White) are dying of AIDS because of their "Brown" skin. In effect, Caucasians have become victims of their own propaganda!!!![36]

This Web site refers to the discovery of the delta 32 mutation, which, it is theorized, enabled some residents of Eyam, England, to survive the 1665 outbreak of the bubonic plague. Using village records of survivors, Stephen O'Brien traced their direct descendants and determined that there is a statistically significant presence of a genetic mutation named CCR5 delta 32. This mutation, researchers theorize, protects its bearers from such diseases as HIV that use the immune system against itself by blocking the pathway to infection. When two copies of this mutation are inherited, the bearer lacks the receptors necessary for the replication of this class of pathogens.

The delta 32 mutation is a trait borne almost exclusively by those descended from northern Europeans, including Americans. As Alison P. Galvani and Montgomery Slatkin affirm, "The high frequency, recent origin, and geographic distribution of the CCR5-Delta 32 deletion allele together indicate that it has been intensely selected in Europe."[37] The allele does not exist at all in African populations, and its incidence in Asia is only 2 percent. Recent studies show that those who bear two copies of delta 32 are apparently immune to HIV, though scientists disagree about which disease, plague or smallpox, the mutation initially provided protection against. Based on his studies of the descendants of Eyam (all of whose ancestors survived the plague), O'Brien concludes that the plague was responsible.

The inequitable distribution of the delta 32 allele and the coincident appearance of HIV-AIDs apparently has given rise to conspiracy theories, such as the one quoted above. The Stewart Synopsis Web site supplies "evidence" that HIV was a genocidal invention of the U.S. government. "About 10 percent of Europeans carry the CCR5 Delta 32-mutation," the site points out. "The incidence is only 2 percent in central Asia, and the mutation is completely absent among East Asians, Africans, and American Indians. THAT SHOULD TELL YOU SOMETHING."

Whatever it might tell us, the site continues with its main thesis: "The HIV/AIDS enzyme is the product of many steps in the laboratory according to all scientific criteria in every independent 'de novo' review that has been conducted to date. The science history shows an 'Aryan obsession' with development of ethnic biological weapons targeting people of Negroid descent. At present it is unclear exactly when the genociders learned there was an exploitable difference in the blood of the Negroid Race."

The point here is neither the paranoia nor the bad science of the site's authors but how information about genetic distributions unwittingly reinforces racial stereotypes that genetic investigations have declared to be nongenetic. The Stewart Synopsis Web site arranges a set of facts: the existence of the delta 32 mutation among those of European descent, the emergence of HIV-AIDS in Africa, and one further notable fact—that J. Craig Venter Jr. holds the patent to the gene called the "African American HIV/AIDS Entryway." This is the same gene that early U.S. scientists used on African children in the late 1950s (CCR5 delta 32 positive)—to unveil the racial conspiratorial origins of HIV.

Which, of course, does not account for its high incidence among gay white men.

In this complicated scenario of racist longings and continued fears, recent genetic science has simply (or really not so simply) come clean. The announcement of the discovery of no genetic difference has complex cultural implications even if the announcement appears to undermine some of them. Less clear are the cultural implications of announcing that humans and chimpanzees share 98 percent of the same genes. One could imagine that this pronouncement helps animal rights activists make their case for the decent treatment of animals. After all, if humans and chimps are that much the same, don't chimps deserve to be treated as well as humans? In fact, if shared genes define some essential identity, doesn't all of life deserve the same treatment? The problem is that declaring genetic similarity is a slightly misleading way to understand the biosphere, even on a genetic level. If we subscribe to a notion that genes are selfish and out for themselves, a biosphere of shared genes is a biosphere full of both conflict and cooperation. Appealing to a species of species kinship does not solve problems of aggression, exploitation, and survival (look at the human nuclear family), nor does it provide the basis for any enlightened ethics, since the fact of shared genes does not alter competing notions of what might constitute a natural justice.

The genetic similarity between humans and chimps, when added to the declared disappearance of racial differences, does invite us to resituate ourselves in relation to the world around us if we believe that genetic codes define us. That resituation, while it may signal an evolution toward a more cosmic line of thought, also provides the context for and necessity that the "truth" of binary gender be reaffirmed. This gender truth is not the biological fact of cocontributing organisms typical of sexual reproduction. Rather, it is the truth of a binary gender imagined in the nostalgic terms of the American nuclear family with specific essential binary gender roles. Ward and June Cleaver become the models of gendered genetic interactions, mooring up the loss of other artificially binary organizations such as race with an imaginary domestic scenario that confers well-being and order into a world in which other orders are rapidly disappearing.

Wishful Thinking: Genes, Longevity, and Immortality

When we imagine genes as agents, they become literal representatives of our bodies, our wills, and our desires. We become our genes and our genes become us, so that we imagine that we, too, somehow, survive from generation to generation. We parse up faces, bodies, and behaviors, attributing a nose to this parent, a gait to that, a disposition to a grandparent, or those bedroom eyes to a bachelor uncle. But wait a minute. A bachelor uncle? The figment of genetic survival conveys two contradictory presuppositions. First, the passage of genetic material constitutes some sort of survival: we somehow survive personally through our genes. This is a version of the connection between genes and ideas of immortality observed by Dorothy Nelkin and M. Susan Lindee in their study of our images of genes. Nelkin and Lindee call this "genetic essentialism," which "reduces the self to a molecular entity, equating human beings, in all their social, historical, and moral complexity, with their genes."[38] But second, even if we do not individually pass on our own genes, parts of us still survive through a familial gene pool. We are immortal if we pass our genes on; we're immortal if we don't, but someone else does.

Both of these ideas of survival relegate some immortalizing function to genes, but do so through an appeal to the connection between genotype and family, an appeal played out culturally in naming practices, property, estate planning, and other signifiers of familial continuity. The confusion between familial and individual legacy through the transmission of genes is the same confusion that produces and sustains genealogical versions of individual identity that already confuses genotypes with identity and personality in the first place. The case of identical twins would show, among other things, that individuals are more than simply the sum of their genes. Common sense would dictate that no matter how hard we try to invest genes with a representative function, the truth of the matter is that there is no personal immortality at all, not even a genetic one. These confusions cast the individual as the product and culmination of a history, which history, as it is selectively condensed into a few phrases, can define the individual. The problem, of course, is that only genes survive—or only their information—and individuals may carry a vast array of genes from multiple sources, increasingly so as people more readily move around and increasingly meaninglessly as genotypes are divorced from cultural contexts. But of course genes always were more or less divorced from culture so that the function of the gene as already a synecdoche of a complex array of cultural ideas, biases, and stereotypes becomes quite clear.

Despite our narrative of a direct line of genetic contribution, we actually believe in family resemblance, a brand of identity that does not require direct genetic contribution, only membership in an identifiable gene confluence defined as an extended family. Within this gene pool, traits become accidental pieces, identified by patronyms, that constitute the illusion of survival and immortality. In the project of family resemblance, one concept of survival is conflated with another. If we imagine genes as agents, then their survival in our children is somehow our own survival. The literal survival of genes becomes the figurative survival of the individual—a figuration actually taken as literal, even though we know that in terms of any individual, immortality is not the case. Biological transmission stands in the place of personal contribution and survival beyond death. Without considering how this can possibly be, the gene comes to stand for both the personal and the biological legacy. But do I feel better about the prospect of death because my nephew has my twenty-three cowlicks? How can they be "my" cowlicks? How is it that phenotypic traits become confused with genotypes that themselves become confused (as our recent spate of forensic DNA identifications would tell us) with individuals? What hope lies here?

Tracing familial traits and resemblances is a complex activity. Such tracing occurs on the basis of phenotypic traits—that is, genetic traits visible in the shape and appearance of bodies or, more impressionistically, in certain behaviors. Because phenotypes are only the tip of the genetic iceberg, tracing the reappearance of phenotypic traits always involves a guessing game about where recessive genes might have ended up. Insofar as this process tracks some correlation between phenotype and genotype—made much more exciting when recessive traits, such as blue eyes, appear in subsequent generations—it is reminiscent of Mendel's first labors with garden peas. But tracing familial genes also expands the scale of racial or ethnic traits where certain phenotypic traits such as the shape of eyelids or noses or even earlobes are read as representative of certain racial types that themselves define the history of a family's alliances. There is, in other words, a way in which phenotypic traits serve as the signs of family history and identity. So, for example, the brown-eyed, dark-complected child who shows up in a family of northern European, Scots-Irish is perceived as reflecting some long-forgotten French or Spanish connection (even though such darkness could just as well be a manifestation of Celtic or even African bloodlines). And the tall, graceful child who appears in a family of shorter, stockier children appears as a reminder of some more aristocratic contribution to the peasant muscularity of immigrant farmers.

In the context of more concerted discernments of racial heredity, the phenotypic markers of "blood" serve less savory political, cultural, and economic ends. In turn, the narratives of genetic contribution in family lines—one's mother may be Scots-Irish and Dutch while one's father is German and English, or one's father is black and one's mother French—constitutes a notion of individual identity within a historical matrix premised on nationalistic stereotypes. This occurs especially in the United States where such ethnic and racial mixes are commonplace, often still traceable, and often significant in attempts to redress historical imbalances and oppressions.

Studying phenotypic similarities among generations of a family indeed certifies, at least insofar as family resemblance can certify, the perpetuation of a family's genes or some of them. Part of the pleasure in having discussions of resemblances comes from contemplating what seems to have survived from parents and grandparents. It also becomes data for identity and identification. A child who has her father's nose and her mother's dimples is assured of legitimacy and belonging. At the same time and more unwittingly, such conversations are attempts to estimate what portions of each parent's genetic material a child received. Because in human reproduction each parent's germ cells' twenty-three half chromosomes are cut and pasted and may be paired with genes from the other parent that are more dominant or recessive than they are, individuals carry both parents' genes but may look different from either one.

More striking differences occur if the combination of parents' genes carries forward two genes that were recessive and unexpressed in each parent. One of the most obvious cases of this is when neither parent has blue eyes (a recessive trait), but the couple gives birth to a blue-eyed child. If both parents carry the recessive gene for blue eyes, the chance of one child having blue eyes—receiving two copies of the recessive gene—is 25 percent. Recessive traits in children, which are precisely not chips off the old block, signal a kind of genetic dis-ease. Genetic disease genes, like the gene for cystic fibrosis, are also often recessive, carried by parents, but not expressing themselves until a child receives a copy of the recessive gene from both parents. The appearance of such genetic diseases signifies that one's genes were never one's own, that parents, too, are merely conveyors for generations before them. If spitting images confer pride in continuity, children with recessive traits remind parents of their parents, of a line longer than themselves. The appearance of recessive traits is also the revelation of a parental "secret." Until a child with recessive traits is born, it is unlikely that a parent knows whether he or she even carries the recessive gene.

The prospect of genetic inheritance—that one's offspring will bear one's traits as an outer sign of the perpetuation of one's genes—retrospectively affords a basis for the other kinds of inheritance: the devolution of wealth and property from parents to children. Legacies of property have long served as symbols of familial connection, particularly in cultures such as classical Rome where the family was the source of social position and power. The idea of an estate that passes to one's heirs provided a means for the orderly transmission of property and goods from one generation to another. But it also provides the economic means to sustain the familial genotype. Marked by position and reputation, familial identity and wealth become synonymous, the first the means by which the second remains intact and useful to a defined group of people. The concept of some sort of biological or even legal inheritance existed long before the discovery of genes as "blood" or "flesh," protected by the father's name as the symbolic seal of legitimacy and belonging. The discovery of gene pools and genetic inheritance merely provides a specific biochemical basis for a mechanism already known to exist and already girded by and itself undergirding law.

Or, rather, we never actually question whether genotypes do parallel and underwrite inheritance. Instead, our understandings of inheritance and legitimacy influence the kinds of questions we might ask about the identity of genes and family in the first place. After all, in any given child a certain number of genes from each parent don't survive. As much as may survive are also lost, and those that we passed on may well be expressed differently in different genetic combinations. Certain combinations of genes halt familial perpetuation, as is the case where children inherit genetic combinations that make them sick, sterile, or indisposed toward reproductive activity. Is this a genetic line in the process of committing suicide? Or has this loss always been a necessary part of survival, as important in its way as perpetuation may be in its? Does the loss of familial reproduction mean that any such genetic line has been selected out, or do the same genes simply proliferate elsewhere and thrive in other circumstances?

Consider, for example, what the advent of catastrophic war does to the chances for survival of male children. If a family has all male children, its chances of surviving are less than if the family also has some girls or all girls. Genes that may not work so well during one kind of cycle—say a cycle of plenty where bearers of genes that make people susceptible to high cholesterol die quickly—may provide an advantage in a cycle of starvation. The point is that our narratives of genetic survival celebrate what appears to survive rather than see the process as a far more complex economy of gain, loss, reserve, and variety. They are also selectively deployed to underwrite cultural organizations such as patriarchy. When genes are conflated with the survival of the father's name, this confusion celebrates male offspring, even though children of either sex can pass genes on in equal measure. Although in patriarchal societies sons are valued because they are more able to perpetuate familial property or may be better situated to help parents in their old age, genetically favoring sons makes little biological sense. Part of the impetus to conflate genes and survival comes from popular narratives of Darwinian evolution that equate survival with the continued ability to reproduce understood in terms of the visibility (the phenotypic appearance) of beneficial traits. Part comes from the idea that in the narrative of reproduction, reproduction itself is the antidote (or more likely the consolation prize) to loss understood as individual death rather than as the loss of half of a genotype.

In fact, our characterizations of genetic agency, particularly the image of genes as little homunculi, reiterate not beliefs about the familial identity of blood but ideas about both the nature of evolution and earlier ideas about the mechanisms of human reproduction—and about how these two processes are ideologically intertwined. In other words, our notions of reproduction itself depend on the idea of the homunculus, and our ideals about evolution depend on our ideas of reproduction. Popular understandings of evolution, therefore and despite evolution's more reasoned science, still depend on a notion of the gene as a willful agent.

The Evolution of Reproduction

From its earliest formulations, reproduction involved the transmission of an agenic representative particle. In the fifth–fourth centuries B.C., Hippocrates developed the theory of "pangenesis" to account for how it happened that parental traits passed to children. Hippocrates' version of pangenesis located sperm, which concentrated elements from throughout the body, as the agent of passage. In addition to carrying particles from all of the organs, semen also carried traits that had been acquired during the life of the semen bearer.[39]

In the mid-fourth century B.C. Aristotle, a student of Hippocrates, disagreed with Hippocrates' notion of pangenesis and introduced his own concept of seminal action in which he imagined that the semen's "vital heat" molded new beings by heating menstrual blood. He disproved pangenesis by showing that individuals who lost limbs did not pass that trait on to their children. The discussion of these two thinkers sets out the parameters of most later discussions of human reproduction: the central role of semen as an agent that carries forward its bearer's traits and the argument around pangenesis that will continue until the later nineteenth century.

William Harvey, the man who discerned how blood circulates in the body, suggested another theory, that of "epigenesis," or the idea that the body's organs form indeterminate globs of matter contained in the germ cells. Germ cells thus contained a protoplasmic version of the individual. A third theory, "Preformationism," emerged in the seventeenth century, ironically encouraged by the recent development of the microscope. Preformationists believed that sperm contained a completely formed tiny individual called a "homunculus." Their ideas were supported by pictures drawn by microscopists who drew images of homunculi-filled sperm as seen under the microscope. Caspar F. Wolff (1734–94) confirmed Harvey's theory in the eighteenth century and disproved preformationism by closely observing the embryonic development of chicken eggs, adding the observation that organs developed from undifferentiated cells. In the eighteenth century Pierre de Maupertuis (1698–1759) reinvented Hippocrates' idea that sperm carried representative particles from around the body. Although Maupertuis believed that sperm were dominant, he conceived a contributory role for the egg so that a new child represented the inmixing of elements from both parents.

By the mid-nineteenth century, Charles Darwin employed a notion of pangenesis to account for how changes accrued in evolution. Like Hippocrates, he conceived of sperm as carrying small particles, which he dubbed "gemmules," and like the French scientist Jean-Baptiste, Chevalier de Lamarck, believed that children could inherit characteristics acquired by their parents. Darwin's cousin Francis Galton tried to prove the mechanism of pangenesis by seeing if traits could pass from rabbit to rabbit through blood transfusions, but found no evidence. By 1863 August Weissman finally disproved the idea that inherited characteristics could be acquired.

These concepts of reproduction were obviously linked to attempts to account for the inheritance of traits. Theories such as Aristotle's and those of the Preformationists understood heterosexual reproduction as the action of form (or information) carried by the sperm on the egg's matter. This reflected ideas of male superiority, but also an inclination toward reproductive material one could see. Adherents of pangenesis might allow for contributions from both parents but, apart from the details of actual pregnancy (how to grow a child), did not actually require the inmixing of information from both parents to pass changing traits from generation to generation. Everyone seemed to know that two parents were necessary to produce a child, but all assigned a more active and interesting role to the male parent, whose contribution was actually more easily questionable (and hence more securely reinforced by social conventions such as naming).

But what made heterosexual reproduction better than the asexual reproduction represented by "lower" forms of life such as the simple splitting of paramecia, the hermaphroditism of earthworms, or the kinds of asexual reproduction employed by bees and ants? Although heterosexual reproduction was a founding truth of biblical religions and the accrued wisdom of human experience, why it was necessary seems to have produced a certain anxiety, as reflected in the work of Sigmund Freud, as the coequal roles of male and female came to light in the nineteenth century and as increasing certainties about paternity, enabled through such devices as blood typing, became available from the beginning of the twentieth century on.[40]

That a concept of "blood" supplies a scheme for the orderly devolution of wealth points toward the rather significant denial represented by the satisfactions of passing something on. Passing on bodily traits or bank accounts is simultaneously the symbol of an extended survival ("I continue to exist in my nephew's earlobes") and the admission of mortality ("If someone else has my TIAA-CREF benefits, I must no longer exist"). Survival becomes confused with immortality; the perpetuation of genes becomes wishfully confused with the perpetuation of life. This confusion is not a real intellectual confusion in that no one actually believes that individual consciousness survives with one's eye color or vacation home. It is, rather, a conceptual entanglement in which the perpetuation of one aspect of self symbolizes some kind of group memory that staves off the absolutism of individual mortality. The confusion of survival and immortality also parallels the paradox of sex and death in which sex itself is envisioned as a precursor to death. In both decedents' estates and genetic legacy the condition of continuance—the passage of wealth or traits—simultaneously signals the inevitable cessation of existence. The logic of this sex-death connection, as explained by the French philosopher Georges Bataille, is that reproduction, imagined as the act of perpetuating self, requires the death of that self as the cost of its perpetuation.[41] Long-standing cultural associations between sex and death persist in narratives of doomed lovers (Romeo and Juliet); in the French slang for orgasm as *la petite mort,* or little death; and in the connections we assume between aging and a loss of sexual vitality.

Just as inheritance represents an ambivalence about survival (I survive, but I am dead as a condition of this survival), so the sex-death connection instills a confusion between sex and reproduction. If reproduction ultimately requires the death of the parent, as is more immediately evident in the reproductive practices of insects, why would nonreproductive sex require the same sacrifice? Is this simply a version of the Garden of Eden myth, where the price of sex is the death of innocence, a mode of restraint imposed by puritanical caveats against pleasure, or does it in fact collapse sex, reproduction, and aging as the inevitable facts of mortality? However the confusion comes about, the link between reproduction and death embodies a tragedy with its own antidote. Death is offset by the production of beings who carry with them pieces of the parental genes, which are now the transmitters of more than biochemical information.

Genes have come to bear the load of individual identity, survival, and denials of mortality, as more figurative and poetic notions of reproduction and legacy bow to the conquering truth of a single set of chemicals. The agency we attribute to genes is an agency required by the psychic and ideological burdens they now carry. Genes are not automatons, we believe, because mere chemistry drains the significance from the fact that we who now exist represent already the survivors of eons of genetic selection. We are the champions of evolution, and we are reflected in and are the reflections of our genes, which have the cutthroat will to triumph, the stuff to come out ahead (or come out at all). Our genes are our partners in the grand adventure of survival and refinement. It is only right and natural that we attribute to them the same traits that got us here.

The End of Evolution As We Know It

We may be creatures of our genes, but we have long imagined that we control them. Breeding practices developed with crops and livestock have made us husbands (an interesting term). Eugenics, the human form of husbandry, is another way humans attempt to control genetic action. Jewish communities where marriages are arranged by matchmakers have been able, for example, to keep track of which community members carried the recessive genes for Tay-Sachs disease (a genetic condition that causes mental retardation, paralysis, and early death). By screening communities for the gene, keeping track of the carriers, and not arranging marriages between them, matchmakers were able to eliminate Tay-Sachs from the population (though the genes themselves persist).

Humans have long practiced a variety of eugenics based on class affiliations to the point that the term for an upper-class individual, at least in Anglo-Saxon countries, is someone who is "well bred." It is usually the case, however, that we understand eugenics as a short-sighted, often racist practice, especially as pursued by the Nazis in Germany, who sterilized those whom they thought "racially impure" and rewarded "Aryan" mothers who had a lot of children. This is because, among many other reasons, eugenics advocates base their ideas of breeding on specious qualities and ideological evaluations that have as much to do with environment, opportunity, and wealth as they are linked in any way to genes or their perpetuation. Fears about eugenics, then, come not only from a sense of potential disenfranchisements based on spurious conclusions about individual value but also from the ways eugenics programs reiterate racial and class biases and dismiss individual interest in a genetic legacy.

The originator of the term *eugenics* (changed to this from the less fortunate "viriculture") was Darwin's cousin Galton, who was an early champion of the heritable quality of behaviors. Thinking that through pangenesis, gemmules for various behaviors and mental habits were passed into the mix, Galton became convinced through his own statistical study of "genius" (eminence in a particular field such as art or wrestling) that there was a greater chance of geniuses having genius sons than the chances of a genius son arising from the general population. Although he did not know about genes or understand exactly how inheritance worked except through the crude postulate of gemmules, Galton produced a study of exactly how many geniuses produced genius sons and the degree to which they were related to other geniuses. As he promises in his introduction to the 1869 study *Hereditary Genius:*

> I propose to show in this book that a man's natural abilities are derived by inheritance, under exactly the same limitations as are the form and physical features of the whole organic world. Consequently, as it is easy, notwithstanding those limitations, to obtain by careful selection a permanent breed of dogs or horses gifted with peculiar powers of running, or of doing anything else, so it would be quite practicable to produce a highly-gifted race of men by judicious marriages during several consecutive generations.[42]

Apart from conceiving of humans as analogous to horses and dogs, Galton's idea of breeding talented men misappreciates the ideological complexity of the idea of "genius," the complexity of the role of the genes in the production of mental attributes, and the degree to which populations prefer to have that selection occur more accidentally. The conceptual problem with eugenics is the apparent intervention of human control, which brings with it a host of limitations and scrutinies far worse than any kind of dictatorial tyranny. Though Galton's motives might have been philanthropic (in the longest-term sense of the word), the assumption of knowledge was hubristic from the start, as Galton decried natural selection as slow and aimless. "I shall show," he says, "that social agencies of an ordinary character, whose influences are little suspected, are at this moment working towards the degradation of human nature, and that others are working toward its improvement."[43] He concludes that "each generation has enormous power over the natural gifts of those that follow, and maintain that it is a duty we owe to humanity to investigate the range of that power, and to exercise it in a way that, without being unwise towards ourselves, shall be most advantageous to future inhabitants of the earth."[44]

Galton's hubris enacts the problematic tautology inherent to most notions of eugenics. Apparently men of genius determine the breeding of geniuses (or the Nazi version, superior races suppress inferior races). In other words, eugenics involves perpetuating the powerful as a kind of narcissistic program of immortality dressed as species improvement. Those in power presumably oversee the production of beings like those in power because the powerful (or the geniuses) are likely to define desirable qualities based on their own obvious superiority (itself based on their being in power). This is the selfish gene theory on a macro level.

Whether we like it or not, we practice eugenics all the time on a social and medical rather than on a "natural" scale. Medical science permits individuals, who a century ago would have perished before reproducing, to survive and reproduce. The increasingly uneven distribution of wealth, population movements, and global industries speed the rate through which previously disparate populations intermarry. There is no overt program or locus of control, and with that absence we imagine that nature takes its course. Or perhaps, as Samuel Beckett's character Clov announces in the play *Endgame,* "There is no more nature." The social, whether we recognize it or not, has already become a large arbiter of selection.

Rewriting the Genome

In the name of life we fear genetic engineering, the site where the metaphor of the text intersects with the imagined power to author. If genes are books and DNA is an alphabet, then what is to stop us from changing the story? We like to think of those changes as beneficial. The concept of gene therapy, for example, imagines the insertion of either the proper or a corrective gene in place of genes that are damaged or whose present version is expressed in a genetic illness such as cystic fibrosis. The idea is that doctors simply rewrite the bad script, and, if imagined through the metaphor of the text, this indeed seems like a simple task. It was so conceptually available, at least to the general public, because of the pervasiveness of the gene/text metaphor that gene therapies were tried in the early 1990s well before we had the capability of really succeeding (if we do now). Early examples of gene therapy were attempts to treat SCID.[45]

SCID, or Severe Combined Immunodeficiency, involves a defect in the lymphocyte (white blood cell) systems of the body. The defect causes white blood cells to disappear, leaving the body vulnerable to infection. Most SCID patients, such as the famous "bubble boy" David, who lived enclosed in a germ-free bubble, succumb to infections as children. The disappearance of white blood cells is caused by one of several different genetic changes. One change occurs on the X chromosome, which affects only males. Another causes a deficiency in ADA (adenosine deaminase). A deficiency in ADA, which is a metabolic enzyme necessary to the maintenance of the immune system, causes the immune system to fail.

Unlike many disorders, where treatment would require intervention into the billions of cells throughout the body, genetically caused SCIDs could theoretically be treated by introducing a corrective gene into T cells and reintroducing the T cells into the bloodstream or by introducing the proper gene into the stem cells of the bone marrow. This kind of intervention is possible because the ADA deficiency is caused by only one mutation (instead of several) and because only a small amount of ADA is necessary to maintain the immune system. Thus, if only a small number of targeted cells could be "infected" with a gene to produce ADA, the deficiency could be managed.

This gene therapy began in 1990, conducted by Michael Blaese of the National Institutes of Health, who took the approach of reintroducing T cells engineered to include the gene that produces ADA into the blood. It was simultaneously pioneered by a research team in Italy that focused on bone marrow. Both approaches were successful, but not without risks and an almost prohibitively high cost. And both therapies were augmented by an enzyme replacement therapy, PEG-DNA (also prohibitively expensive). Gene therapies for other disorders used mouse retro viruses (engineered to remove the harmful genes and to include the helpful ones) to introduce beneficial genes into the body.[46]

We still assume that gene therapy will succeed after the kinks are worked out, but we imagine that it will work for more common conditions such as cancer, heart disease, Parkinson's disease, and sickle cell anemia. We also, however, imagine a more invasive type of genetic engineering in which developing embryos are invaded with genes designed to enhance capabilities. While this version of genetic engineering also depends on the analogy of language, we fear this one, since in many ways it looks like a version of eugenics. Like the eugenicist, the would-be genetic engineer improver meddles with nature, loosing unknown ills through the apparently innocent alteration of a single genetic element. Or so we imagine. This genetic engineer creates a superrace of beings who are resistant, strong, brilliant, athletic. Or so we imagine in such films as *Gattaca*. What we fear is the caste war that ensues. What we fear is some genetic twist that irrevocably destroys the gene pool by introducing some runaway virus, like a computer virus, that replicates everything in its own image, among other things.

But, as Stephen Hawking prophesizes in *The Universe in a Nutshell*, genetic engineering is inevitable.[47] We will learn how to do it. Using the analogy of informational bits, Hawking traces the acceleration of information accumulation from the slow changes accumulated in DNA representing millions of years to the suddenly increasing accumulations produced in the last century or so, not only in the biological record of DNA but in other forms such as printed texts and computer files. Biological record turns quite literally in Hawking's view to the written and electronic record. After all, it is all information. Despite the fact that there is a difference between these kinds of information—DNA information, recall, is performative, that is, it begins a cascade of processes, while most textual and electronic information is merely record without any intrinsic performative effect (apart perhaps from software)—Hawking predicts the inevitable acceleration to the point of genetic engineering. What is to prevent one kind of information from marrying the other? Perhaps Hawking is right, if not in his use of the information analogy, then in his understanding of the psychology of progress that masks capitalist interest. It is certainly the case that chemical companies sell engineered crop seed designed both to resist the company's herbicides and to not reproduce so that farmers cannot produce their own seed.

And so what if we even get to the point where we can engineer humans? Apart from the ethical dilemmas that accompany all attempts to mess with what is perceived as the natural order of things, the effect will be to alter evolution forever. If we select and alter, there will no longer be any natural selection—or any natural selection that might occur will occur in relation to a very different set of circumstances that are no longer simply environmental. If, for example, as molecular scientists (and Dawkins) suggest, genes compete with one another even within the body, what happens to competition when a spate of alien genes is introduced? The competition will certainly change, as will processes of selection. Selection will no longer be natural, fitness will no longer depend strictly on environment, and evolution as we know it will end.

Clearly, this picture, too, is merely an extension of the fears about control shared with our dislike for eugenics. But, like eugenics, we might want to consider the idea that evolution as we know it has already ended. Reproductive fitness, for example, is no longer defined strictly in relation to the environment anyway. Selection is no longer natural. Technologies have insulated us from the kinds of conditions that exact biological change. But they may spur others to the point where human evolution occurs in tandem with technology. In this sense of interdependency we become cyberbeings, as Donna Haraway has shown.[48]

Our fears of engineering are linked, thus, to our other fears about genetic information: who controls it and how will they use it. The most prominent ethical issue discussed in popular literature devoted to issues about genetics (such as the ELSI section of the Human Genome Project) is the question of how genetic information is to be disseminated and used.[49] Like most discussions of the potentials of genes, this one, too, is premature, since no one really knows exactly how the genome turns into the body. But the problem with DNA is that whether or not we understand it, companies and the government might use it anyway in the form of a bank of DNA profiles of criminal offenders, for example, or insurance companies requiring DNA profiles before insuring. To what extent, ethicists ask, should our DNA stand for us?

We need to be careful about our uses of DNA as information, since we are never looking strictly at DNA but at the plethora of associations, ideologies, metaphors, and narratives that reengage conventional and often outmoded ways of thinking. We might think a DNA profile is quite an advanced idea, the ultimate mark of individuality. But think about it. DNA is a profile of statistical probability. Like fingerprints, we think it unique; it can even deposit itself far more liberally and inadvertently. If we think of DNA as the chemically inscribed version of a person, aren't we merely translating the accidents of heredity into identity? Aren't we making ourselves conceptually into books to be read even though the genome is not a book and we can't read it anyway? How is this finally any different than the use of patronyms or place-names to identify individuals? Other than our belief in DNA's utter facticity and uniqueness, what is different about it and the ways we have always thought individuals to be unique?

In its figurative capacities, DNA, conflated with the idea of the gene, has taken over older metaphoric functions such as the name. It consolidates competing notions of individuality into a code that reduces idiosyncrasy to the statistical probability of combinations. This is all done in the name of science, but it is also a cultural process of transformation from one kind of technology and way of thinking to another, a transformation displaced onto the DNA gene as actor and agent.

CHAPTER 5

The Ecstasies of Pseudoscience

Currently in American popular culture, DNA's subtending narrative of heteroreproduction is being joined or even supplanted by another, less literally reproductive but more oedipal version of that narrative: the question of origins. More the primal scene than the sex scene, the narrative of origins (Where do I come from? Who am I? Who are you?) tends to elide cause and effect chains in favor of a more simple, direct, nominalistic answer that links DNA to an individual without any intermediary. Instead of a logic of production that merges process with agency, origins evokes the practice of naming—of giving bodies meaning by appending a signifier. In these narratives, identity matched with a name is the answer. The beginning merged with the end provides completion. Think, for example, of how Oedipus's fate was prophesied. The play only reaffirmed what the beginning told us, that he was the one he sought.

DNA figures as an alternative for the name, which, though importing a metonymic logic instead of the name's metaphor, nonetheless comes to fulfill the same metaphoric function as the name. Not only, then, does DNA become the answer to the question of origins (and inscribe the future); in doing so it again conserves an older, more metaphoric logic while harboring a potential shift to something else. Because the narrative of origins bypasses causality, it is a version of magical thinking, focused on the inherent power of the word. Thus, in addition to the distortions already appending through its various figurations, DNA also becomes the answer to origins, by which any notion of complex cause and effect or system is supplanted in favor of instantaneity, immediacy, magical associations, and ultimately pseudoscience.

This DNA answer to the question of who one is emerges not because of parental squeamishness in the matter of procreative description but as the effect both of the nature of the question at a time when the idea of the subject itself is unstable and the logical end of the ever-iterated dynamic of the heteroreproduction narrative itself. The question of origins transforms immediately and as if magically into a question of identity, not only because the cause-effect chain itself is elided but because we already understand origins to be identity: we understand that narratively, the end—identity—is always in the beginning, in some notion of origins. The question of identity reorganizes the links between kinship, the reproductive narrative, and conceptions about identity producing the illusion that the question is already answered by either parentage or geographic and ethnic locations. Because DNA functions as the protagonist or figurative parent in the reproductive narrative and because we understand ourselves to be the end products of that narrative, DNA becomes the answer to this other question as well.

Even asking "who" is not asking "how." The story of origins erases process in favor of a metaphor that takes the form of DNA understood as a tag or "real" name. The elision of process also synecdochizes DNA into a singular, stable, even economic formulation that transforms it even further from an agent of biochemical replication to the almost sole representative figure of both individuality and identity. This synecdochizing process bypasses science altogether in favor of a signifying instantaneity that substitutes magical thinking and pseudoscientific alternatives in the place of elided narratives and processes. Reproductive narratives of connection linked to evolution become essentialized identities grounded in the artificial specter of a DNA rendered absolutely nominal, a bar-code answering definitively issues of identity and origins and certifying individuality in the face of ideas of the split subject, subjects intersected and defined by information, and global commodification that in their way have already shifted our notion of who we are anyway.

From Blood to Genes

Nominalized, pseudoscientific DNA instantiates the heredity narratives incident to our saga of reproduction. Fraught with sex/death ambivalence, issues of relation and connection are the flip side of the story—the perpetuation of an identity understood as contiguous and individually defining. Before DNA, blood was the synecdoche that supported this story, which seems to need some literal "substance" to substantiate it, a substance that by its very role and position takes on functions and values and a significance it doesn't otherwise have.

Aristotle thought blood was the nutrient that fed life. Pliny the Elder (23–79) thought blood was life itself. Jewish philosophers thought that blood was the soul. These notions of blood as the substance that animates, houses the soul, and is the essence of an individual continued through the eighteenth century. In the latter half of the eighteenth century, the experimentalist John Hunter (1728–93) called blood a "life principle" that accounted for the differences between animate and inanimate matter. Because blood changed color, because it circulated, because it leaked and could be tapped, blood took on a symbolic function in relation to understandings of the anatomy. Blood, in other words, rarely stayed blood, but also became something else: food, life, soul.[1]

Although no one (at least no one whose ideas were taken seriously) seems to have thought that blood itself was the agent that transmitted bits of parents to offspring, blood has nonetheless become a lasting synecdoche for kinship, a way to conceive the connections among fathers, mothers, and children through generations—what we now understand to be genetic relationships. Blood figures what we imagine has been transmitted. We have bloodlines and blood relatives; we share blood. All of these signify what we now consider to be some shared genetic inheritance to which we give great significance. Blood is thicker than water. Adopted children assert a "right" to know their bloodlines.

Blood as kinship metaphor is also extended to more metaphorically familial groups: tribes, nations, races. In all of these uses, blood becomes a more symbolic figuration of a relationship like and possibly derived from some sense of physiological affiliation. Blood's alliances extend into and constitute the basis for various legal formulations: the devolution of wealth, the title to property (such as fee tail male), the formation of fiefdoms and nations, limitations on intermarriage based on consanguinity, and miscegenation.[2] If blood was life, then families shared the same life. If there was too much blood, then damage could happen. If the blood was too alien, then its inmixture would alienate a bloodline from itself.

Blood usually worked as a condition turned to metaphor that stood for a set of physical connections no one understood, but which figured as an important determinant in social relations, the assignment of privilege, and the structure of societies. The ease with which blood performs this fairly vast set of metaphoric functions suggests that, like genes, there is some cross-cultural need to provide a figure—in this case, a synecdoche—for a range of often unknown processes. The figure for life becomes a representative for more metaphysical properties. The shift from physical substance to spiritual medium encompasses notions of kinship and identity, the signature of individual life, the qualities inherited, and the individual's place in the socius. It also signals how substances can, in their transfiguration, come to ground magical thinking. The question is whether they are transfigured in the first place as an effect of this urge toward immediate connection and elision of process that characterizes magical thinking, or whether their availability as figures elicits this mode of thinking, or both.

Blood became a literal means for individual identification after 1900 when Karl Landsteiner discovered blood antigen groups—O, A, B, AB—that could help determine whose blood was compatible with whose for transfusions. More symbolically, this typing of blood could help determine who might not be the father of a child.[3] Contemporaneous with Landsteiner's discovery of blood groups, William Bateson, an Oxford geneticist, translated Mendel's work on heredity, more or less ignored since the 1880s. Though there were no connections between the two—

Bateson's readers were not thinking about the connection between Mendel's factors and Landsteiner's blood groups—the beginning demystification of blood and incipient explorations into genetics overlap. Blood, although still a favorite metaphor for kinship and heredity today, soon began to relinquish some of its metaphoric functions to the idea of the gene. Not only does blood transfer its figuration of family relations to the image of genetic connection, but genetic connections become a synecdoche of the larger set of social relations previously identified as "blood" relations. This slow process has taken place throughout the twentieth century until we are at the point where we usually understand that the properties of blood are somehow the effects of genetic determination. We might, therefore, understand the metaphor of blood as a lingering precursor to the role of the gene after the mid-twentieth century. The gene, which we think of as fact instead of as metaphor, represents a different, yet still imaginary, register through which we begin to conceive of and express sets of filiation and relation as evidentiary and traceable.

Blood's complex significations, however, are a lot for genes to bear, especially as genes' reference to a specific set of interlinked processes also signals a shift from what have been blood's largely metaphoric relations to a conception of the world as intricate, interconnected, and causally chained. What blood previously represented was a series of relations and conditions characterized only by the ubiquity of blood's presence. Although various thinkers hypothesized mechanisms for how blood animated the body (it carried nutrients from the intestines, it was "heated" in the heart or lungs, it carried heat throughout the body), there was no literal connection between blood and the supposed mechanisms of reproduction, or any visual similarity between the blood of various members of families—except that they all had it, and it all looked alike.

As DNA takes over the figurative functions of blood, DNA genes, already metaphorized, become even more so. Just as blood substitutes for a series of unknown or misunderstood processes, so genes (and DNA) supposedly make such relations as identity and filiation certain. DNA spells it out, so to speak. Although DNA is conceived as a set of facts—

in fact, the base-level set of facts—its status as factual character itself works as a figurative expression of the same larger concepts as blood. DNA equals life. DNA proves relation. DNA establishes identity.

The most significant difference between the metaphor of blood and the figurative workings of genes and DNA is in the ways we imagine that their functions are accomplished. On the one hand, blood circulates and spills. It stands in for but provides no mechanism to explain a series of complex and even metaphysical concepts. On the other hand, genes and DNA are imagined as having some physical connection to the processes they govern, even if most people cannot imagine what those processes are. If blood is a substitute—a metaphor—then genes and DNA are imagined as contiguous to or metonymic: a literal, physical part of the larger process they govern. Genes, thus, represent a slightly different mode of conceiving how a single element might stand in for a complex set of processes and relationships. In taking over the figurative functions of blood, DNA genes signal simultaneously a shift from a mode of substitution (or metaphor) to a more mechanical (metonymic) concept of order. In so doing, they conserve metaphoric and paradigmatic structural thinking as they poise for a transformation to systems.

What seems, in fact, like a slight shift in figurative language represents a much more crucial but gradual shift in the logic through which we conceive the order of things: the difference, for example, between thinking of the father's name as certifying (even producing) a relation we couldn't prove and thinking of DNA as a way to certify parental identity. The first logic is metaphoric, where a word is substituted for an uncertain relation. The second is metonymic, where physical phenomena are linked to one another and where such practices as naming have a basis in some provable chain.

The DNA gene introduces a different fantasy for ordering relations as well as for understanding the mechanisms for how concepts stand for other less known processes. But as a synecdoche, DNA is also the figurative heir to blood, the final answer to the oedipal son's question of identity. It is both the harbinger of a different, more metonymic order and the conservator of older ways of thinking about body, identity, kinship, connection, and perpetuation. DNA may answer Oedipus's question in a different way, but it is still made to address the question—in fact, it seems to reaffirm the significance and facticity of the question itself in what appears to be the ultimate performance of a reductionist answer that finally brings everything (body, name, trace) together in what is the smallest piece of life of all.

The DNA gene seems to offer the possibility of a complete account of how living organisms get from molecules to intelligent life. The possibility of linking the minute evidence of connection to larger phenomena—another version of the origin-identity story—animates recent preoccupations with forensic science and DNA evidence that arguably produce a completely interconnected trail from deed to perpetrator: a real chain of evidence, so to speak. The DNA synecdoche also enables imaginations of genetic bases for all behaviors, instincts, and even desires. This idea of a substantive chaining from DNA to protein to the basic bioprocesses of body maintenance to behavior offers a logic of immediate causality based on contiguity—of leaping from one element to the one beside it as a way to make sense of phenomena. The problem is that the chain is still incomplete, at least with the DNA gene. Criminalists might be able to forge a "chain" of evidence, but genetic scientists still do not know exactly how the genome gets to be the living body. This place of unknowing is one of the pressures that incites the use of such metaphors as textuality and agency. But this unknowing also transforms the DNA gene itself into a metaphor for almost the very same processes blood stood for: life, animation, identity, kinship.

Just as blood became a metaphor for the metaphysical, so the DNA gene threatens to become a metaphor for the same vague, phantasmic properties. Blood was the best guess of two millennia of thinkers as scientists; it carried the pretense of science. The DNA gene, now the best guess of a century of thinkers and scientists, unfortunately also threatens to become the pretense of science as it is offered as the explanation of everything without regard to how it works and what it can do. A willingness to assume connection provides the elision of the actual mechanisms of connection in favor of immediacy of a fractal nature. This delusion appears when we think that the processes themselves, infinitely repeating, need not be filled in. The enthusiastic boasting of

Francis Crick in 1953—"We've found the secret of life"—turns into the totem for this assumed connection that feeds a cultural appetite for pseudoscience. Like the metaphors deployed to explain genes, this pseudoscience offers a magical, fictional explanation of how genes work in place of the one that is messy or overly complex. This fiction reinstalls a fantasy of human control that genes threaten to destroy by superseding human will and agency. But it also offers a return to the worldview in which the relation between a substance and a behavior, like the four humors, was conceptual, metaphoric, and easily malleable.

So at the moment in history when we have enough incipient insight into the possible mechanisms of life to explore the complexities of life processes, we take the elements of that knowledge—DNA, genes—and transform them into symbols through which the familiar, empowering magic of pseudoscience returns. The process is compensatory in the same ways metaphors such as the book of life and blueprint are—they give us the illusion of control over processes beyond our control. But pseudoscience also offers another advantage: it offers immediate, transcendental action instead of the very slow processes of mutation and evolution. It substitutes will and desire for mechanisms of survival. Or it substitutes a notion of "intelligent design" for the accidents of evolution, a notion that retrieves a place for deities in the pantheon of science.[4] We think that it tells us not only who we are but where we came from and how we are unique.

A Rhetoric of Pseudoscience

Bypassing process, establishing DNA as an efficacious signifier, brings together questions of biochemical process, origin, and identity.[5] Treating DNA as if it actually were a language (enabled by removing its processes in the first place) enables a series of cultural practices and justifications for those practices that not only are pseudoscience but also conserve and justify a sense of human ascendancy, subjective and cultural certainty, corporate profit, and even science's efficacies. As the Word, as a powerful, magic signifier, DNA becomes available to almost any account of the universe, the meaning of things, or the nature of life.

As Carl Sagan has observed and as we all more or less know, science and most religions offer differing understandings of life, its appearance and development, and the future of living things.[6] Religions usually base their accounts on divinely inspired messages that have become scriptures. These scriptures can be read literally (usually known as fundamentalism) or more figuratively in which events or aphorisms become the rich basis for relevant interpretations. The relation between the "word" and belief, word and ethical behavior, is almost always a matter of interpreting a text that has, because of its sheer age, become somewhat figurative for most people. Understanding sacred texts as figurative is a way to solve the problem of eternal relevance. If a sacred text was written during an age when there was no electricity or internal combustion engine, what the text might say about light or travel must be reinterpreted in light of contemporary possibilities.

This habit of reconciling divinely inspired advice with contemporary material conditions and knowledge has extended itself to the insights of science that would seem to clash directly with the universal schemes mapped in sacred texts. How, for example, do we align Darwin's notion of evolution or theories of the big bang with a biblical or many folk renditions of creation? How do we understand the problem of evil in a culture that tries to explain behaviors scientifically? What is life when we have the capacity to forge it outside the womb and extend it sometimes almost indefinitely on machines?

Many religions translate. They see, for example, the days of creation as eons figuratively expressed. They understand Satan as psychiatric disorders or genetic propensities. The deity's will is expressed in the way things go, and religious luminaries can find cause-effect meaning in the coincidence of almost any event. The process is a poetics—the interpretation and rerendition of events or ideas from one system of knowledge into another. But having rendered DNA a signifier already makes it very easy to move between these disparate discourses—and others.

The translation of science into magical thinking involves several figurative mechanisms that derive from classical rhetoric. Via the metaphor of language and the elision of process, these rhetorical devices enact a series of transpositions that are then displaced back onto a DNA synecdoche. These devices are simultaneously the strategies of both the popularization of DNA genes and pseudoscience in general. The rhetorical mechanisms include the following:

1. Progressive displacement in the substitution of one cause for another. The causes often offered are understood as basic or root causes—prime movers, so to speak. For example, we previously thought that the "Devil" made someone do something evil. Now we think that a bad environment, bad parenting, or mental illness produce bad behavior. All of these "causes" are metaphors in that they substitute for one another to offer a cause for the same behavior. None of them alone actually traces any chain of causality—how the Devil or bad parenting get to be manifestations of bad behavior. Such causes stand in for an entire chain of cause and effect of which, at best, they are only a part.

2. Metathesis and allegory in the transposition of events into sacred narratives or myth. This process takes events that are more or less literal—bad behavior, earthquakes, a birthmark—and matches them to a preexisting pattern or paradigm. For example, bad behavior signals satanic influence (especially if the culprit has a birthmark), or if a series of massive earthquakes occurs, some prophet will point out that such events are the beginnings of Armageddon, or tragedy of any sort will be declared God's punishment.

3. Metalepsis in the leap around chains of causality to link a prime cause to a prime effect. If a magician utters a magic word and a rabbit appears, the utterance seems to produce the rabbit. The series of machinations by which the rabbit was produced are eliminated from the account.

4. Metaphor in the substitution of an entirely different set of cause-effect relations organized according to the conventional logic of narrative and aimed at accounting in culturally familiar ways for the cause-effect leap (rabbit from hat because of abracadabra). We rarely try to spell out the series of actions that produce the illusion of a magically appearing rabbit, but we do like to air cause-effect accounts for other kinds of "magic." For example, we think tooth decay is caused by sugar (itself already a metalepsis as in 3 above). We substitute "Mr. Tooth Decay" for the sugar. Then we produce a narrative of victorious conquest—Toothpaste strong-arms Mr. Tooth Decay away—for the chemical processes

by which the bacterial action that causes tooth decay is retarded by debris removal and fluoride. The complex dental interaction becomes a simple story of good versus evil located in the battling personas of a top-hatted malefactor and a gleaming white hero.

Through this rhetoric, DNA genes become the pseudoscientific agents of a series of fantasy narratives of genetic causality, curative ability, and overall determinative power. What has become clear in the past fifty years is that DNA is anything but a simple molecule and that genes work through the complex interactions of systems rather than as independent agents with a direct causal relation to any given effect. The pseudoscientific erasure of this systemic complexity may be simply a matter of public relations with the resistant or undereducated, or a matter of strategic characterization for continued funding and support, or the inevitable effect of representation itself. But however it happens, it also results in an understanding of DNA genes as pseudoscience. Only the miraculous operations of the pseudo still ironically provoke the kind of faith and credibility we might wish people would reserve for real science. The appearance of magic—instantaneity, power, singular cause and effect—is the register of empowerment in a rapidly technologized culture, a register admirably fulfilled by other mechanisms that erase or mask all traces of their operation such as digital computers or complex electronics.

The Abracadabra of DNA Genes

There seems always to have been a propensity to render DNA genes in pseudoscientific terms, at least in hyperbole and synecdoche. From their very beginnings as a public substance, DNA genes have been endowed with a kind of magical power—the "secret" of life—the signifier of most biological causation. DNA genes are imagined to be responsible for every aspect of an organism, its development, and its behaviors. The simple statement that "genes make us who we are" while implying a cause and effect relation actually proposes a magical pseudoscientific operation—something like a cake recipe or perhaps more accurately (since cakes are themselves complex) something like encapsulated sponge animals that take their shape when the capsule is dissolved in water. We don't know how genes make us who we are, we simply now endow something we call a gene with a sort of one-on-one orchestrative power. In this sense, from its first description as a molecule, the DNA gene has functioned as a synecdoche that stands for a complex and intricate process.

The process for which the gene itself substituted, however, quickly became not a process of complex biochemical interactions but one more like a combination lock. If one dials or punches the correct numbers in the correct order, a door opens, or in the case of a DNA gene, we somehow imagine that knowing the order of nucleotides somehow opens the door to knowledge and control. It is as if DNA itself were the magic word that makes life materialize from nothing. Biochemistry becomes an incantation accompanying the prestidigitation that produced life. Like the very word *abracadabra,* whose lack of referent (what is an abracadabra?) signifies its function as the pure power to produce or transform, so the three letters of DNA became referentless signifiers of transformation and production.[7] And, like abracadabra, the mechanisms of DNA operation are hidden, mysterious, and occult.

If there is any mechanism attributed to DNA in public culture, it is like abracadabra, the magic word. The magical link between DNA genes and language transposes the intricate relations of accident, selection, and multiple, interlinked processes into something like writing and reading. Writing is the inscription of meaning into symbols. Reading is the translation of those symbols. Both writing and reading involve a transformation from one form (idea) to another (symbol) that can be reversed. The actual chain of processes by which idea becomes symbol is usually elided in favor of a narrative of writing as a one-on-one process of intent, meaning, and control as opposed to biochemistry's (or the fledgling writer's) sense of accident, complexity, and indirection. DNA, as embodied and reduced to its linguistic metaphors, operates as a signifier with stable reference and material effect. It becomes literally Austin's "speech act."

The expressive allegory of DNA is, thus, also linked to the biblical sense of the "Word" as immediately causal. The coincidence of the idea of DNA as the foundational or primary substance (where "word" meets deed) and the biblical Word as equally primary and foundational associates the two. This correspondence inserts DNA into a large but somewhat inapt narrative and belief system as the biochemical agent of a more allegorical process by which the will of a deity, represented as the Word, becomes physically manifest. In like manner such a process inserts a deity into the workings of DNA.

This particular stage of transposition, or metalepsis, is evident in the idea of "intelligent design" proposed by scientists such as Michael Behe in *Darwin's Black Box*. Behe and others who wish to salvage the role of a deity in molecular biology and evolution proffer the argument that some biochemical processes, such as blood clotting, constitute "irreducibly complex systems." In such a system, according to Behe, which is "composed of several interacting parts that contribute to the basic function, . . . the removal of any one of the parts causes the system to effectively stop functioning."[8] This kind of system, we might agree, seems fairly pervasive in biochemistry or (as a point of comparison) in such machines as the internal combustion engine, the computer, or even a coffeemaker. Behe goes further, however, to assert that such irreducibly complex systems "can not be produced gradually by slight, successive modifications of a precursor system, since any precursor to an irreducibly complex system is by definition non-functional."[9]

What Behe is claiming is that in an evolutionary scenario, an irreducibly complex system could not have emerged gradually via natural selection because any of the single components of that system would not have functioned to perform the given purpose (i.e., blood clotting) without the whole system, nor could there be any system to perform the function without all of the parts organized in a complex sequence. Thus there was no basis for the gradual developments or improvements of evolution. How could blood clotting evolve as blood clotting if it couldn't be blood clotting all at once? If a complex system such as blood clotting couldn't have emerged from evolution, then it must have come about in some other way. The difference between an irreducibly complex biochemical system and a coffeemaker is, presumably, the imaginary presence of an organizing intelligence. Behe's claim would be that like the coffeemaker, blood clotting has an engineer. In the case of biochemistry, that engineer is a deity.

This kind of argument is a negative argument, with all of the negative's logical flaws—if we can't find it, it must not be there. Blood clotting is too complicated to have developed through evolution, so something else must have produced it. This kind of negative argument (apart from working as a confession of a contemporary failure of imagination) is enabled precisely by the unwitting transposition of the DNA gene into a performative signifier. Behe's argument begins with the premise that some processes can simply be willed into existence. To this end, the biblical narrative of the Word conveniently bypasses all causal chains in favor of the mysterious will of omnipotence. Behe's inability to produce an account for irreducibly complex processes through the mechanisms of evolution produces a circumstance in which it appears that there can be no cause-effect chain. The absence of this causal chain, which also characterizes the narrative of deific creation, creates an analogy between divine creation and evolution, which is attractive for other more ideological reasons. The apparent lack of a cause-effect chain then becomes (metastasizes) into the deific narrative where cause and effect disappear in favor of the efficacy of deific will: "Let there be light." Let there be blood clotting.

Whether or not we believe in the creative powers of deities, "Let there be light" is a version of abracadabra. What Behe is saying is not so much that somewhere there is a large brain intelligently designing irreducibly complex processes but that somewhere there is a deity saying abracadabra. Other scientists, of course, dismiss Behe's claims as nonscience, showing that indeed the apparent "chasm" of evolutionary process can be explained. Behe's argument does serve other ends in addition to being a symptomatic example of the rhetorical processes of a transposition into pseudoscience caught in midstream. Obviously, Behe's appeal to a deity and location of holes in evolution enables religion to resolve the apparent clash between scriptural and scientific explanations of life. In some communities in the United States, such magical thinking is credited as a legitimate alternative to actual science, a misconception that pits science against pseudoscience as a matter of public will and educational policy.

The notion of intelligent design represents another step in DNA's transposition to pseudoscience by eliminating even the allegorical causal chain provided by the metaphor of language. Obviously, appeals to religion remove such causal chains immediately. But in a more secular register, the allegory of language itself removes cause and effect equally expeditiously (though not as doctrinally), since we all know how language works. Or especially because we never really think about how language works. The complex cause-effect chain of molecular chemistry becomes a matter of expression and implied will. DNA renders life as if DNA's code were transparent, as if code and meaning shuffled equivocally. You equal your DNA. Your DNA equals you. The effects of communication and expression become the intent of the DNA gene (recall the easy slide from the notion of the selfish gene to an imputation of genetic intent). In this way the gene as a prime cause is linked to all biology as its effect, but without any sense of the complex steps of its relation to other molecules. DNA is also credited with phenomena that are not the direct and simple result of DNA alone, such as behaviors or cancers. Simplicity comes to stand for complexity. Abracadabra produces a rabbit from a hat. This is pseudoscience.

"At the heart of pseudo-science," Sagan observes, "is the idea that wishing makes it so."[10] While science requires the production of testable hypotheses, gathering of evidence in as unbiased a fashion as possible, skepticism, and the ability to alter hypotheses to fit the evidence, pseudoscience proffers untestable hypotheses, garners anecdotal and untestable evidence, can never account for a complete causal chain, and prefers the myth over any evidence to the contrary. Sagan suggests that the popularity of pseudoscience comes precisely from the sense of empowerment provided by wishful thinking. But even this sense of empowerment is no simple phenomenon but a complicated mixture of delusion and ideology that compensates for the very lack of individual empowerment science seems always to impart. This isn't science's fault but is rather an effect of science on a culture in which individual power and personal significance have long been crucial fantasies. Scientific explanations suggest a rather impersonal set of mechanisms governing life and the universe. Most ideologies work toward organizing and sustaining systems of empowerment based on one criterion or another (wealth, gender, race, religion). Science, when it is not the victim of them, generally assails these organizations. And even if science doesn't fall victim to ideology, popularizations of science will.

Through this somewhat complicated rhetorical process, the DNA gene, which seems to be the ultimate in scientific explanations offered to the public, becomes the quintessential example of pseudoscience. Via DNA, magic becomes science. This pseudoscientization has several effects. First, since DNA seems to be accessible science and DNA has become magic, all science becomes magic. The effect is not to increase interest in science as a powerful mode of reasoning (since this science has no reason) but to equate scientific and apocryphal phenomena (parapsychological phenomena, space aliens, crystals, astrology, etc.). A logic of verbal power and totemic immediacy substitutes for the precepts of real science. This has the effect of slowing public acceptance of real science while most unfortunately occupying the space where real scientific reason should be. Hence the confusion Sagan observes between scientifically explainable phenomena such as aurora borealis and plant metabolism and folk myths such as Roswell, Atlantis, and crystal power situated as truly scientific but conspiratorially kept hidden. Real science, this trend would suggest, is so powerful it must be suppressed.

Second, pseudoscience enables false or unreasonable expectations. If DNA works like a magician, it would seem that magical DNA cures should be just around the corner—as soon as we find the formula. While this ensures funding for research and a certain optimism on the part of an aging America, it also instills false hope and a lack of appreciation for the utter complexity of such possibilities. This was particularly true in the first bloom of genetic sophistication in the 1980s and 1990s. By 2003 there were twenty-seven different gene therapies in use, all working on the same basic idea of introducing a corrected or missing copy of a gene into target cells. However, in 2003 the FDA placed these therapies on "hold" because patients receiving gene therapy treatment in France had developed leukemia. What researchers discovered was that retrovirus vectors inserted their material at different sites in genes rather than simply drop off or insert the material in a more random fashion. The monkey retrovirus often used in treatments deposited its information at the beginnings of genes. This has a greater chance of affecting how genes work and was causing some to go wrong. What scientists discerned ten years after the first experimental gene therapies was that genes are not Tinkertoys to be added and subtracted at will but systems whose parts interrelate in a far more complex fashion.[11]

Thus if we think of genes as letters in an alphabet and imagine genetic therapies as a simple matter of correcting spelling, then we also have particular expectations about the possibilities of genetic therapy in general. In an alphabetic logic (where one letter replaces another), genetic therapy should be about as easy as replacing a muffler. This has everything to do with eliding the entire immensely complicated and systemic mechanisms of where genes are, how they work, how they interact with other genes, how biochemistry works in general, and so forth, not to mention the vast numbers of chromosomes in the body. The Human Genome Project—the idea that we now have a complete list of the nucleic acids that make up the "human genome"—reinforces this kind of thinking by telling people that we have the entire "alphabet." This encourages people to expect an increase in the kinds of possible interventions that can be made. After all, if we can spell, we can write. And rewrite.

Third, the pseudoscientization of DNA has its most unfortunate effects perhaps in considerations of ethics where startling misinformation and misapprehension ground what should be reasoned discussions of policy. What happens when we do "rewrite," which is one way we think about engineered agricultural products? Monsanto, for example, introduced a gene into soybeans and cotton that rendered the plants immune to the effects of Roundup™, Monsanto's trademark herbicide. The idea was to produce a "Roundup™-Ready" crop that would enable the use of the herbicide on all of the weeds around it without killing the plant. Clearly this is a commercial idea, not one that necessarily betters the soybean, especially since farmers using the seed needed to sign a contract with Monsanto averring that they will only use Roundup™ on their crops.[12] Apart from the ethical questions about tinkering with the genome for commercial benefit, the economic consequences of Monsanto's policies, and the environmental effects of so much herbicide, the practice raised larger questions about the safety of such tinkering in general. The problem doesn't seem so much to lie in the fact of tinkering but in the perception that evolution and the ecosystem themselves have been tampered with. In other words, we can play with the alphabet all we want, but we better not mess with the larger story.

The larger story, of course, is the ecosystem, a concept that finally introduces a systemic view into popular considerations of genetic science. Science confronts the pseudoscience of corporate promotion. It is not a coincidence that a systemic notion emerges to combat commercial practice, since opponents of commercial biotechnologies savvily recognize the potential systemic effects of any alterations in the planetary genome. Commercial genetic engineering (used widely on nonhuman subjects) includes such practices as giving genetically engineered hormones to cows to increase their milk production, splicing genes for vitamin A into rice, adding a protein gene from the Brazil nut to soybeans, or engineering bacteria to accomplish certain tasks such as "eating" oil to clean up oil spills or transforming waste into ethanol more quickly.

The promotional material deployed both to advertise and defend bioengineering practices relies on the pseudoscience of an alphabetic concept of genes as a way to deflect concerns about safety. If genes are like letters, adding or altering one or two won't hurt the sentence and might improve it. Or, we can't misspell a word if we think of genetic material as undifferentiated "plasm." Monsanto's corporate Web site, for example, calls genetic material "germplasm," of which they claim they have a "collection." "Scientists are now capable," they boast, "of identifying the genes that are responsible for some of these desired traits and are able to manipulate them." The genes or the traits? It doesn't matter. Monsanto's presentation equates the two in a Tinkertoy (or here more of a Mr. Potato Head) notion of the gene/trait as something like an accessory, something one manipulates depending on the circumstances. These interchangeable gene plasmic parts produce, via what Monsanto calls "genetic engineering or genetic modification," "high-quality brand-name seeds." Engineering or modification (whatever the differences are between these two terms) produces "brand-names." Working like detachable parts, Monsanto's plasms produce individualized, identifiable, proprietary seeds with the brand-names that signify their status as property—the logical end to perceiving DNA genes as an alphabet whose different orderings can be owned.

The trouble with commercial applications of biotechnology occurs, however, systemically, and only a systemic view—the kind of view that disappears with pseudoscience—will reveal problems. The problem, for example, of giving genetically engineered growth hormones (rBGH—Monsanto again) to cows to make them lactate is that such overlactation increases the incidence of udder infections, which in turn requires treatment with antibiotics. Both the rBGH and the antibiotics end up in the milk and then in the general population, where their damage has not been calculated. With the philosophy that dumping sewage into the sea creates such a small impact that it is meaningless, biotech companies dump growth hormones and antibiotics into the food chain. The potential effects of this dumping are on cows and human consumers who may, over time, develop problems that have not been anticipated (or at least advertised). Producers who use these hormones are not even required to warn consumers of their presence under the same dumping philosophy, this time employed by the government.[13]

Evidence of the pervasiveness of nonsystemic pseudoscientific conceptions of the genome persists even with the government, whose mandated tests of genetically engineered products stop short of examining the larger systemic realm such changes might affect. The government, too, seems convinced that the genome is an alphabet with interchangeable parts. The systemic approach employed by opponents of widely broadcast engineered genes often finds effects in places that range beyond the more myopic or alphabetic view of the government and corporate enterprise. For example, testing of the Brazil nut–engineered soybean beyond the parameters required by the federal government revealed that the engineered product also contained a dangerous allergen. More extensive testing than required also revealed a serious systemic problem with a bacterium engineered to make it more rapidly change waste into ethanol. In the environment, the bacterium also destroys a soil fungus necessary to enable plants to absorb nitrogen. Had the bacterium been released into the ecosystem, it would have killed fields of plants and rendered the soil sterile.[14]

Another worry of ecosystem advocates is the possibility that engineered genes will escape via pollination into the larger ecosystem, affecting natural species. The basis of this fear is a more Frankensteinian fear that hubristic meddling with nature will have irreversible effects. These fears have been the foundation of horror movies and are premised on an idea that the systemic character of the ecosystem is so complicated that we do not know or cannot anticipate potential effects. In addition, the idea of meddling with evolution produces anxieties that we know not what we do. Of course, we don't, but we do know enough to know that the system is immensely complex, interlocked, and delicately balanced. Certainly, loosing new genetic combinations into the biosphere ends evolution as we know it, but we have already been doing that in other ways—through selective breeding, habitat reduction, and altering the conditions for survival and fitness.

A continued evocation of the pseudoscientific alphabetic concept of the ecogenome is only in the interest of commercial profit despite corporate insistence on the philanthropic motives of its endeavors. Monsanto confesses: "We're excited about the potential for genetically modified food to contribute to a better environment and a sustainable, plentiful, and healthy food supply."[15] In the following sentence, however, this British Web site tries to deflect concerns by suggesting that doubts exist primarily in the minds of consumers: "We recognize, however, that many consumers have genuine concerns about food biotechnology and its impact on their families." The concerns are "genuine" but curiously narcissistic. While Monsanto is thinking in large terms like the "food supply," consumers with doubts think in terms of "its impact on their families." Not, notice, about biotechnology's impact on the environment, or the genome, or evolution. Monsanto thinks of the world, consumers think of themselves.

But, in a slogan that reveals just how open-minded Monsanto is, the corporation declares: "Food Biotechnology—is a matter of opinions. Monsanto believes you should hear all of them." Apparently, food biotechnology is not a science, though in terms of its presentation to the public we already knew that. Public science is simply a matter of opinions. Biotechnology, according to Monsanto, apparently means simply hearing these different opinions. Though superficially this appears to be an appeal for open-mindedness, we know what this means. The scientific arguments that opponents of genetic engineered foods raise are merely opinion. In a war of opinions, would the public prefer the conservative naysaying of "scientific" opinion or the farsighted philanthropic largesse of corporate beneficence?

The war between biotechnology corporations and advocates of the ecosystem is more than a battle over opinions—over whether improvements in the quantity and quality of engineered food crops outweigh largely unseen potential effects on the biosphere or human health. It is a war of concepts, the same war outlined throughout this book between structural ideas of the DNA gene—the alphabetic, transcribable, ownable code—and a systemic view of completely interlinked, interdependent parts and processes. The ideological stake of this battle, finally, may boil down to profit. The structural view enables commercialization. A systemic view warns against such practices without extensive testing. This is not a matter of forward-looking companies versus conservative environmentalists but a matter of the very basic conception of how life on earth works, a matter of pseudoscience versus science.

The brand-names Monsanto bruits about point to the fourth danger of thinking about DNA pseudoscientifically—as an efficacious magic word. If DNA genes are seen as letters or interchangeable parts that magically produce certain traits or effects, then brand-names, premised on some specific engineered genetic combination, perform a quadruple identificatory function. First, they identify or represent a particular genetic combination (though usually referring really only to one or two added genes) or product. Second, they imply an ownership of the combination, which, to be owned, must be unique and reliably the same. Third, they identify the owner. Fourth, they point to the simple fact that the seed has been altered. We are familiar with the fact that agribusiness has been breeding crop seed for years; we have seen their signs and experimental fields as we drive down highways. The difficulty with brand-named genetically "modified" seed is that it is presented in the same way as other "naturally" bred seed products. We cannot tell from the fact of a brand-name alone whether seed has been bred or modified, which suggests a kind of corporate bait and switch, or an "official" attitude pretending there is no difference at all. As agricultural products reach the American table, there is no requirement that the consumer ever be told whether food has been genetically modified.

The brand-name serves multiple commercial purposes and completes the transformation of biomatter into property and product. With human beings, DNA's link to naming is reversed. Instead of naming genetic combinations for the purposes of trademark, we imagine that human DNA combinations will serve to identify their carriers. Our DNA becomes our trademark, substituting for names, numbers, or any other individualizing criteria. Endowing DNA with the capacity to "name," or identify, individuals encourages a culture of surveillance and paranoia, especially in our somewhat schizophrenic belief in both the unique character of individual genotypes and the sweeping knowledge of the human genome (which is no one's genotype), in our belief that our genome is ours and the simultaneous granting of property rights in identified genes as information. Our genes are ours. Knowledge about them belongs to someone else. Genes are specific; the genome constitutes a bank of knowledge. Somehow, we believe, we can get from the one to the other in a process that works kind of like a menu. One (or nature) selects certain options or variations, and the particular combination chosen then represents some unique expression out of the vast set of possibilities offered by the genome. All of this is based on the statistical probability that the DNA of any two individuals will not represent the same choice of all the possible variables out of a vast "library" of options. Except in the case of identical twins.

Profiles

Statistical improbability, thus, underwrites the instant connection between the body and its DNA name. This instant connection stands in for an assumed trail of linked elements that lead from origin to identity and grounds the forensic uses of DNA, but the probable nature of that probability is covered over by our belief in the agency of the name. Identifications made on the basis of DNA are produced via a DNA "profile," sometimes inaccurately referred to as a DNA "fingerprint." The comparison of DNA with fingerprints, made because modes of DNA identification have aspirations to the accuracy of fingerprints, is evoked to suggest both the certain uniqueness of the DNA profile—like a fingerprint—and its forensic use to ascertain that a particular body was at a particular place or handled particular objects. The term *profiling* already bears a certain connotation of probability rather than certainty. But it also suggests a surveillance and scientifism that make profiling activities creditable, acceptable modes of control. Profiling is law by probability.

Actual fingerprinting, however, is not a matter of probability at all but a matter of near certainty, practiced in a primitive but insistent reminder of physicality. The practice of fingerprinting grew from a desire to link someone who signed a contract with the contract or promise he had made. Sir William Herschel, a magistrate in colonial India during the 1800s, required natives to make a handprint on contracts they signed. Although Herschel's motive was more a magical one—to impress on signers the importance of their promise through a laying on of "hands"—Herschel did notice that the handprints could be linked to the individuals who had made them. Herschel's accidental discovery of the identificatory powers of handprints was at the time merely another footnote to the series of observations about finger ridges, discovered by the Renaissance microscopist Marcello Malpighi (after whom a layer of skin was named). John Purkinji, a professor of anatomy, observed nine different classes of fingerprints early in the nineteenth century.[16]

Fingerprint lore existed side by side with nineteenth-century efforts to identify repeat criminals. Having eschewed more primitive methods of identification such as branding, tattooing, or mayhem, criminologists relied on the photographic memories of officers. In the 1870s Alphonse Bertillon, an anthropologist, tried to devise a formula for recording the dimensions of certain bones, thought not to change in adults. Called the "Bertillon System," the measurement practice was used through the late nineteenth century until a pair of identical twins with the same name who were both criminals showed the flaw in the system.

It wasn't until 1880 that a British colonial surgeon in a Tokyo hospital, Dr. Henry Faulds, devised a classification system for fingerprints, which he, too, had noticed seemed unique. Faulds sent his system to the elderly Charles Darwin, who passed the information along to Francis Galton. Faulds did recognize the potential of fingerprints for identification, devising the method of impressing them with ink. Galton, who had been the recipient of Faulds's material, published a book on fingerprints in 1892. While Faulds (and curiously Mark Twain in *Life on the Mississippi*) was more interested in fingerprint identification as an aid to solving crimes, Galton, as might be expected, was interested in seeing if fingerprints yielded identifiable patterns of race and heredity. In other words, within fifty years, fingerprints had evolved from physical marks to synecdoches of identity with surplus meaning.

Galton, in fact, originally thought that fingerprints would work, like stereotypical phenotypic traits, to enable him to trace kinship and racial patterns. In his 1888 publication "Personal Identification and Description," Galton concludes, "I should say that one of the inducements to making these inquiries into personal identification has been to discover independent features suitable for hereditary investigation. It has been my hope, though utterly without direct experimental corroboration thus far, that if a considerable number of variable and independent features could be catalogued, it might be possible to trace kinship with considerable certainty."[17]

Galton did settle for the potentials of fingerprints as a mode of personal identification, though not without a curiously colonialist bent:

> Some of the latest specimens that I have seen are by Mr. Gilbert Thomson, an officer of the American Geological Survey, who, being in Arizona, and having to make his orders for payment on a camp suttler, hit upon the expedient of using his own thumb-mark to serve the same purpose as the elaborate scroll engrave on blank cheques—namely, to make the alteration of figures written on it, impossible without detection. I possess copies of two of his cheques. A San Francisco photographer, Mr. Tabor, made enlarged photographs of the finger-marks of Chinese, and his proposal seems to have been seriously considered as a means of identifying Chinese immigrants. I may say that I can obtain no verification of a common statement that the method is in actual use in the prisons of China. The thumb-mark has been used there as elsewhere in

> attestation of deeds, much as a man might make an impression with a common seal, not his own, and say, "This is my act and deed"; but I cannot hear of any elaborate system of finger-marks having ever been employed in China for the identification of prisoners. It was however, largely used in India, by Sir William Herschel, twenty-eight years ago, when he was an officer of the Bengal Civil Service. He found it to be most successful in preventing personation, and in putting an end to disputes about the authenticity of deeds.[18]

Fingerprinting as a mode of certification came from China but was then employed to inveigle colonial subjects to stick by their contracts. The notion of the fingerprint as a species of seal—of an act that guarantees the identity and sincerity of the signer—begins the connection between fingerprint and identity. Turning the guarantee around to use the fingerprint to identify unknown perpetrators makes the guarantee of presence into the ability to discern presence and guarantee the accuracy of the identification. It should be noted that even fingerprint identification, like DNA, is based on a series of comparison points and the likelihood that more than one individual would have specific combinations of such points.

Fingerprints have worked as a way to discern presence—and by presence I mean the intersection, at some point of time, between a finger and an object on which the finger left the print. The arrangement of ridges on the fingers, palms, and feet is unique to each individual, including identical twins (or so it seems). The pattern does not seem to change through growth or aging. Fingerprints can be left on almost any object through the almost invisible deposit of body fats, perspiration, or other substances that collect on the fingers. They would, thus, seem to be a regularly appearing and reliable means of identification. The problem with fingerprints is not their reliability but, for forensic purposes, the ways malefactors have found to avoid leaving them.

A DNA "fingerprint" seems to solve that problem, at least temporarily. People inevitably, constantly, and inadvertently shed bodily substances containing DNA: saliva, sweat, epithelial (skin) cells, hairs, blood, fingernails. All of these artifacts, if processed properly and if not contaminated by substances that would alter their DNA, can yield a DNA profile. All parts of us are us, we are in all—pangenesis yet again, as DNA becomes an ineffable synecdoche of individuality and identity. The idea of a DNA profile is based on the fact that the DNA of each individual differs slightly. Some areas of genes vary widely and can contain many different combinations called "polymorphisms." One such area has a series of repeated nucleotides, called short tandem repeats (STRs). These might be something like "agtagtagt," where the combination "agt" is repeated three times (or four or ten). Selecting for these STR sites yields a variety of alternatives in the number and makeup of the repeated nucleotides.

Although each person's genome is immense, forensic scientists can narrow the sites on DNA samples they test to a manageable number of STRs. Selecting STR sites sufficient for testing provides a series of variations in highly variable areas. If forensic scientists always test for the same sites and compare the results in a uniform fashion, then chances are they will find a unique set of variations for each individual. If they have data on the probability of the appearance of each variable in each population, then they can offer the statistical probability for the appearance of each variable. They thus can produce a "profile" of a panel of variations with the statistical likelihood of the appearance of each. And the panel or profile can be stored as identifying information in a computer to be searched for a match when necessary.

The FBI uses what is called a CODIS STR system. This system employs thirteen STR sites, or "loci," that consist of chains of four repeated elements (e.g., "aaggaagg").[19] The variables of the repeated sequences for each site are analyzed according to the statistical probability that the variable will appear in the population. Accumulating thirteen such variables, each with a statistical probability, adds up to an immense probability that no other DNA sample will have the same set of variables. Thus, the reasoning goes, each profile represents a uniquely identifiable individual.

The FBI keeps these profiles in computers for comparison and identification or "matching," although the uniqueness of each profile is a matter of probability rather than a matter of certainty. Matching DNA profiles produces a possibility. This is in keeping with the entire practice of criminal profiling, where sets of characteristics—often behavioral, racial, geographic, and economic—present the likelihood of some malfeasance or sketch a behavioral portrait of a violent criminal. Law enforcement increasingly operates on probability as a way to arrive at certainty. But such probability matching also threatens to curtail the civil rights of individuals who are doing nothing wrong but who simply match the profile. In this sense the practice of profiling is a kind of overkill, tolerated best when the probabilities are the greatest, or most "scientific"—as with DNA.

Forensic uses of DNA have become a surprisingly popular trope for television dramas about criminology. Their rendition on documentary case histories such as *The New Detectives, Forensic Files,* or fictionalized narratives of crime investigation like the proliferations of *CSI* is all about the final and certain securing of an oedipal story. Each instance of DNA identification comes in the form of a little drama—the drama not so coincidentally of Oedipus. Who did it? How will we know? But instead of the conventional series of dilatory interviews with witnesses, technicians process microscopic "trace" evidence. Instead of relying on the pieced manufacture of a history whose terms of coherence inevitably point to the culprit, investigators find the actual physical links between the deed and the doer. Did he leave his DNA? Here it is. When was he here? Here is his footprint clearly made after the idling car left oil on the pavement. How did he do it? Here are the traces of poison in the backwash of the drinking fountain (if there can be backwash on a drinking fountain). And his fingerprints and his epithelials, hair, an eyelash, dropped carelessly behind.

Although detective stories have pervaded popular consciousness for almost two centuries, providing the satisfactions of solution, the fascination with the genre was as much for the mastery represented by solution as it was for the complex entanglements of ratiocination and tracking performed by a Sherlock Holmes, a Monsieur Dupin, or an Ellery Queen. Contemporary forensic narratives, though still a part of the detective genre, shift their emphasis to matching profile to profile, or linking trace evidence to its source. Ratiocination becomes resourceful technology, tracking becomes collecting. Evidence turns from motive to trace. Or to motivation motivated by trace. The scene of detection moves to a crime scene literalized through its wealth of traces and to the laboratory, which becomes the site of interpretation, thus turning mystery into "science" or the appearance thereof. Through this veneer of science, DNA and other trace evidence represents a fulfillment of narrative promise in complete certainty. We will know the answer, and we will know it because we will have actually established a physical chain from the DNA of the "perp" to the crime itself. The chain is forged, complete, unassailable (except in some notable cases like O. J. Simpson where the jury rejected the probative value of DNA evidence introduced at trial based on doubts raised by accusations of investigators' racism). Who did it? DNA will tell all, because in the end DNA is already the story by which the question of "who?" is answered.

The satisfaction in these stories lies in their certain connection of word and deed, mystery and uncontestable solution—a solution that most crucially is not about motive, psychology, or the vagaries of fate or coincidence that come later as support. Forensic narratives are always about the certainties of physical connection and our new superiority in being able to navigate intelligently through the micro world where we have always relegated the source of meaning or the answers to questions about the physical world. And although these narratives seem oedipal in their impetus, they actually sidestep Oedipus's story to present the assurance that for once, unlike our centuries of repeating Oedipus's task of finding himself out as the killer, the killer is someone else. This shift from Oedipus's paradox of self-knowledge to the nonparadoxical exteriorization of cause represents a significant shift in both our concepts of the individual and our very understandings of how stories themselves should go. If in the Oedipus story we are always looking for ourselves, in the forensic tale we become the source of knowledge who seeks to complete the circle of attachments by which knowledge is made whole. Where the Oedipus story is a drama about individual responsibility, driven by irony, the forensic narrative is driven by a nonironic dynamic of mastery. The only individual responsible is someone else who is caught not by his conscience but by linking pieces into a chain of contiguous presences in time. Both the oedipal and the forensic are about knowledge; sadly, the forensic narrative has abandoned the layered complexities of irony (we know he knows but doesn't know, etc.) for the chained and networked assurances of connection, of turning probabilities into certainties without a shred of self-doubt.

With its new forensic gloss, the detection narrative has become far simpler, far less character-focused, far more bent on clicking the pieces into place and on the magical ability of technicians to derive and provide DNA matches that lead directly to the answer. These forensic dramas would seem to take the processes of DNA profiling apart to show how DNA identifications involve complex procedures. But instead of showing the processes of extraction and electrophoresis, fictional renditions work to render DNA identifications even more magical and instant—a matter of something like fingerprinting. Even fingerprinting was laborious before computer data banks and matching software automatized the work of finding a match. And perhaps, paradoxically, forensic television seems to be all about collecting and matching evidence, but the processes it makes visible actually hide far more complex processes and concepts that enable forensic techniques. The forensic has become very much like the ways it is rendered as a televisual subject: the seams, gaps, mistakes have been eclipsed in favor of the illusions of a smooth, seamless, whole picture whose rapidity occurs partly because of the collaboration of traces and computers.

If DNA provides the illusion of identification, computers provide the illusion of instantaneity. Together, computers and DNA matching present an operational portrait of immediacy that eclipses any processing time—any time between question and answer. In addition, television renderings of these processes cut processing time further through the use of temporal ellipses. The effect is that DNA queries seem to be answerable in the time of a commercial break. The time it takes for forensic shows to match DNA, even if there is an understood gap in time, is so rapid as to seem immediate. Ask a DNA question, get a DNA answer. It's magic.

The cultural saturation of these newer forensic narratives represents a repetition that is itself suspicious. What is our pleasure in this immediacy, this easy mastery underwritten by the impression of certainties (which are really probabilities)? One answer, of course, is that these renderings of DNA represent a comfortable masking of science: pseudoscience is a far more satisfying alternative, mastery more pleasurable than uncertainty. DNA mastery provides the illusion that we finally have the answer, at least to questions of human identity. The world fits together, and there are no gaps in knowledge. Science, with its uncertainties and constant shifting of truths, is over. DNA is the word. Second, such renderings shift science from hypotheses, skepticism, and experimentation to a matter of technology operated (almost uniformly in fictional narrative at least) by unreconstructed sexy geeks. With technology (as opposed to science) any notion of question or mystery has been reduced from the frightening unknown to a matter of feeding data into a machine. All science is hidden by the facades and screens of computers and other laboratory machinery, which become themselves the primary representation of science. Finally, however, our satisfaction comes with the illusion of a set of processes, which, albeit mostly hidden, always work. The forensic drama is the functioning answer machine providing the fascinating spectacle of its reliable functioning. Further, this forensic machine is never wrong—only the human institutions that operate or hamper it make mistakes. Pseudoscience takes on the guise of science in science's own increasingly technologized, machinic guise. Analysis is a matter of input and printout—there is no work involved. Processes are covered by instant answers; detritus sanitized by shiny machines, low lighting, and bloodless bodies; forensic labs look more like spaceships than laboratories, more like alchemy than investigation.

Screening

The overkill of profiling becomes potentially more dangerous—more injurious to individuals with less certainty of any social good—with suggested practices of DNA screening. One kind of screening involves the mass analysis of individuals to find a particular DNA profile. If we know the DNA profile of a criminal, we could conceivably ask any number of individuals based on some vague association with the victim or locale to contribute DNA samples for analysis. This kind of needle-in-a-haystack screening clearly violates individuals' Fourth Amendment right not to be unreasonably searched. But in an era when criminal profiling constitutes probable cause for a stop and search, screening expeditions may not be that far behind.

But why? The same kind of thinking that admits DNA probabilities as proof of identity underwrites all notions of profiling and screening with even less probability of a match. Our faith in the forensic has helped our acceptance of this, which points to the somewhat dubious ideological function of forensic television for promoting the acceptance of probabilities as a matter of a technology that is always certain. If we begin to conduct "scientific" law enforcement according to sets of statistical probabilities presented as truth, we have lost science itself. If we produce identities as a function of probabilities, the individual, too, has disappeared, replaced by a set of chances that stand in for some illusion of scriptural uniqueness. Chances do not represent individuals; they are not fingerprints. Thinking of DNA as a scientific mode of identification and as coterminous with an individual enables the slide from fact and provable deed to polymorphisms and statistical possibilities. This might be an interesting advance if in fact it accompanied a notion of the individual itself as a chatoyant, changing project in process. But that is not the case. Instead, DNA's probabilities work to secure a traditional notion of a stable individuality, one that can always be relied on to point to the same body, through time and even after death. In this way identity becomes sutured to the body as identity's guarantee and repository. And this is a problem insofar as linking identities to bodies has long served as a way to discriminate based on the projections culture makes onto those bodies in terms of value, ability, and privilege.

Seemingly more benign and caring practices of screening, such as the suggested DNA screening of infants at birth for the likelihood of certain diseases such as cancer, heart disease, and hereditary illnesses also cause consternation and for the same reasons as most profiling. Heralded as painless and simple, DNA screening, at least according to the BBC, could provide "parents and doctors with a relatively clear vision of how their [the babies'] health will develop in the years ahead."[20] Although this is a vision of future planning (British health ministers have asked the Human Genetics Commission to look into it), it, too, relies on a pseudoscientific understanding of DNA as operating in a direct and magical relation with certain effects. There is no cancer gene or heart disease gene, and thinking there is misleads us about the value of such an endeavor as well as the simplicity and inevitability of such maladies. Thinking in terms of genes as if they were identification bracelets masks the fact that diseases are generally caused by complex interactions among genes, development, environments, and behaviors, which multiply and complicate the probabilities that conditions will arise. The danger of such screening is that like DNA profiles, they will provide an even less reliable probability of the development of certain conditions throughout the course of a life. But, thinking we can read this old book of life, we will treat certain genetic signifiers as if they are the disease itself. Not only do probabilities decrease, but the length of time any individual remains under scrutiny is increased. DNA screening becomes the equivalent of a perverse Calvinist predestination.

The big worry is that such screening will become the demand of insurance companies, ventures that already operate on the basis of probabilities. And if insurance companies, reading the simple DNA code of a newborn, see heart disease in the future, they can simply eliminate such diseases from their coverage. And if that is the case, why not simply screen blastulas and prevent the gestation of faulty individuals? Or clean up the environmental and social conditions that encourage the development of disease? Of course, because there is not one individual genotype that does not have some potential for disease, we will all be uninsurable. The point is that the absurdity of this forward-thinking "science" is not about improving lives but about a pseudoscientific belief in a magical DNA that exists in a one-on-one alphabetic relationship to diseases and traits and identities. The fact is that no DNA ever presents anything but a chance.

Another concern, as explored by the film *Gattaca*, is that genotypes will provide the basis for "fact-based" discriminations, dependent on the same illusion of developmental certainty as evidenced by genotypes. *Gattaca* shows how the human spirit is not delimited by genotype—how genes are not all. But our desire to know the future based on our very partial readings of nucleotides does represent an anxiety about individual possibility, not in relation to genotype but in the context of a world whose possibilities are rapidly closing down for economic reasons. Fears of genetic typing are really fears about class typing in an era when classes are becoming increasingly separated, and possibilities through education and hard work are disappearing.

DNA genetic technologies, ownerships, therapies, though the products of sophisticated scientific analyses, appear culturally as forms of pseudoscientific magic characterized by certainty, immediacy, and the glossy streamlined covers of analyzing machines. This is science as magic, the conflation of the two enacting finally some illusion of certainty about the very things—the individual, disease—about which we are still uncertain. Somehow in all of this science, we have lost science and replaced it with the image of the code—the perfect emblem of this whole process. We believe DNA is ultimately the name, and that name is the answer to the question of identity. But, as a name, DNA represents something else in terms that must be analyzed. In other words, the name is there both as the answer and as a mystery. Because we believe the name means something (or everything) we take the name as the end of inquiry at the same time that most of us have no idea what the DNA means or how it works. There we have it: science and pseudoscience, answer and mystery. The perfect ignorance in the assurance of knowledge.

Pseudoscience may well be the inevitable effect of any attempt to represent complex processes in language. But the particular character and impetus of this narrative of identity as it has been insistently linked to DNA also tells us something about how DNA works as an allaying trope, compensating and displacing cultural anxieties about mastery, identity, and centrality through our encoding of it as a singular and instantaneous answer. For this reason, DNA reveals not only the problems of representation but how these representations respond to other very different cultural forces and anxieties. Pseudoscience is not, then, merely a goofy fantasy but a complex response to threats to order and alien systems and ways of thinking that are themselves, perhaps paradoxically, contained in the promise of DNA.

CHAPTER 6

Rewriting History

The pseudoscientific capabilities that have attached to the DNA gene—will and agency, the magical elision of cause and effect, instantaneity, the operative power of the Word—derive from the analogies and narratives through which it has been transliterated as well as ways of thinking that constantly reprocess systemic complexity into structural simplicity. Imagining willful agency, while displacing human motives, attributes, and biases onto chemicals, situates DNA as our minute causal agent, as us displaced into it. Envisioning DNA as a tiny entity, which in itself contains and elides all cause-effect mechanisms while also expressing and perpetrating condensed versions of us, renders DNA a tiny little information package, a self-contained kit that tells us who we are, conserves our history, promotes our legacy, produces us, and contains the as-yet-undeciphered instructions for combating disease and mortality. Its combination of instruction and operation, its portability, its shelf life, its ability to individuate and even personalize make DNA the perfect commodity—or the perfect version of an imaginary entity that in itself embodies a shift in our ideas of history, identity, commodities, and commodity systems.

History and identity are revised through appeals to kinship belonging and a version of what we might call "pretermortality," or the idea of having always existed. Several Web sites, for example, now offer ancestry tracing services. A consumer sends (mostly) his DNA to the lab, which compares markers on the Y chromosome to the profiles of a bank of world profiles. Because the Y chromosome is passed only from fathers to sons and differentiates from other Y chromosomes only through the slow accumulation of mutations, those with information banks about the distribution of Y chromosome mutations can estimate from when any particular Y chromosome might derive. The meaning of any results depends on how well information banks have tracked population movement and distribution (or how well theories of such movement play out through data). Information is gained only from comparing markers with those of others, and the results suggest only that two bearers of a similar Y chromosome may have a common ancestor. At best the whole process shows the statistical chance that a given bearer is linked to another bearer.

Such sites are selling identity cast as history. "Trace your roots," one site offers, "Unearth your ancestry. Become a part of history."[1] Since everyone is always already a part of history, what this discourse is really selling is knowledge, or at least an estimation that contributes to an evolving notion of identity as itself genetic. "Create a legacy," it suggests, "that you can share with future generations." Uncannily paralleling what we already think DNA does, Genetrack Biolabs' commodity redoubles DNA by selling its transcription. Another site, Relative Genetics, offers a similar service couched in similar terms: "Discover new family lines. Verify existing family history. Break through your genealogy barriers." (What barriers?)[2] The Genographic Project, which promotes itself as "the first genealogy derived DNA testing service," offers a battery of tests in addition to the standard Y chromosome matching, many based on specific groups such as "African," "Native American," or "Jewish" ancestry tests. The Genographic Project also offers testing of mitochondrial DNA for the ladies and will let you know through comparisons in its database (which now contains your contribution) whether some long-lost relative has recently also inquired about who he or she is.[3]

These sites are selling a commodified instant understanding of identity as the readable link to both history (as itself discerned through DNA) and to a "Family Tree" of all humanity, as the corporations and foundations themselves use specimens to gather data. Through DNA testing we all get to rejoin what we already belonged to. The only difference, apparently, is now we know how more or less, at least in the approximate statistical terms offered by the probability of the appearance of certain markers on one chromosome (and a short one at that) out of forty-six—and that only for males.

The implication of all of these Web sites (and there are at least seven on the Web at this writing) is that we have all become world orphans, detached, lost, divorced from a continuity and history we all long for. Reconnecting with our ancestors will provide definitive identities in place of postmodern subjective splits and identity crises. Science will solve our problems by telling us who we are based on who our ancestors, statistically speaking, were. In the United States, Canada, and the United Kingdom, this is a tantalizing offer for some reason. Certainly for largely immigrant populations, it might solve some of the mysteries produced by ancestors too ignorant, illiterate, or busy to record and preserve any notion of family. But it also offers one notion of identity in place of another. If we have come to think that identity is work—is uncertain, difficult, changing—knowing our ancestors and becoming full-fledged members of the human family tree gives us certainty in place of uncertainty, something objective, traceable, and immutable in place of something winsome, fleeting, and moody.

But what does this genetic relation, assuming it is readable, really tell us about identity or even history? For women, whose link can be traced only via mitochondrial DNA, it doesn't tell us much more than what we probably already know. For men, the Y chromosome comparison offers the possibility of being a Cohen or, better yet, a descendent of the epically potent Niall of Ireland.[4] But what does that mean? Does this suggest that we are now to define ourselves according to our ancestors? And what does that mean?

Dangerously, it might mean the reintroduction (or reification) of ethnic stereotypes. While some regard this kind of project as a proud reclamation of a lost history—see, for example, the recent PBS project *African-American Lives,* which traces both gene markers and documented history to discern the migratory pasts of famous contemporary African Americans—for others it reduces to a recipe of stereotypes (and not even that, since the Y chromosome won't reveal any information about maternal lines).[5] The patchwork of derivations that characterizes the pedigree of most Americans would excuse almost any behavior and at the same time reaffirm behavioral stereotypes as confirmed by history. If someone has ancestors who emigrated from County Tyrone, Würtemberg, Huguenot France, and the Levy's of London, does that mean that such an individual would be a dry-witted, fascist zealot with economical habits? Might such a product not also be an orderly intelligent dipsomaniac who likes gardening and the color orange? And what about that Appalachian interlude? Can we bypass more local, less genetic derivations in favor of a more noble history or select only certain of our forebears to celebrate and totemize? (Preferably the distant ones onto whom we can displace all kinds of assumptions we already have about history.) Such implicit stereotypes threaten, again via a kind of smuggled set of occulted narratives, to substitute essentialist understandings of identity for more complex, situational ones. DNA "history" inserts the "facts" of inheritance in place of much less certain and far more complex psychological processes of development in context. Hidden in all of this are also implicit assumptions about the power that genes have in defining behaviors. The Web sites' answer to the question of identity is the likelihood of genetic relation to a more authentically "ethnic" past. What identity does that produce?

This notion of genetic belonging also reflects a shift in our concept of history. History, which, in the most literal terms, does not exist in the present, is now made to appear materially in the present, carried with us always. When DNA is equated with history, history becomes the answer to any question of origins—in fact, tends to be reduced to the answer of any question about origins. In this way, history also becomes almost completely causal. We are what we are because they were who they were. The present has no will of its own—it is prewritten, prescripted, predetermined. Oedipus is true.

At the same time, history also becomes a tabula rasa for the inscription of contemporary imaginations about previous eras and ethnic character as well as an opportunity to valorize the rugged individualism (and sometimes other less savory behaviors) that enabled our ancestors to survive. We inherited that. And so did everyone else. All of our ancestors were fertile. If, as was proclaimed at the 2000 celebration of the Human Genome Project, we have found humanity's history book, what kind of history does it tell? Is this a history linked inexorably to evolution, the movement of populations, the passages of biochemistry? Or is this a history that is, via DNA itself, always located within us as the key to our mysteries as well as life itself?

The idea, finally, that we represent the culmination of a long line of others whose genetic material we bear in our bodies projects a version of immortality backward. If we have always existed, then might not that pretermortality indicate immortality? Obviously this does not work literally, but as an imaginary continuum, its notion of proven continuity is comforting. And our position as the culmination, the end point also, via a false narrative of progress, renders us the most "advanced" of all. If only our ancestors knew what would eventually flower from their trees.

In having our DNA tested to find out where we come from (whatever that means), we think we are buying our past when what we are buying is a frantic hold on older ways of thinking. There is meaning; it is in our genes. This notion of history as materialized and banked in DNA returns to ideas of DNA genes as performative as well as to the implicit narratives of agency, gender, and control that come with it. The way figurations of DNA genes simultaneously elide systems and even cause and effect and conserve older ways of thinking suggests that the urgency of situating DNA in these ways has much to do with preserving what these figurations perpetuate in the face of something different. The description and evocation of DNA since the mid-twentieth century represents the emergence of a stunningly simple mechanism for negotiating complex modes of thought, primarily the large categories of structure and the emerging theories of systems. This process of defense and compensation occurs across discourses of science, history, identity, and individuality, and goes something like this:

1. DNA comes to embody instantaneous efficacy—of a collapse of signification into direct meaning where the word or signifier is in fact also the signified (performativity).
2. DNA's performative character is not about the truth of the functioning of deoxyribonucleic acid: it reflects the short-circuit of a particular (re)productive narrative whose middling struggles between gendered opposites are excised in favor of an instant link between beginning cause and end effect. We might call this instantaneity, as some historians of biology do, a particular epiphanic leap over means to an end.
3. The ensuing illusion of agenic potency not only makes life processes look superficially simple, its immediacy hides operating sets of complex nonbinary logics and processes such as, for example, the hormonal or cybernetic systems while implying the subtending structural binaries of the apparent cause-effect narrative it has preempted.
4. DNA thus masks with one-on-one immediacy the systems logics that present departures from the binary impetus of the structural story, permitting multiple nonoppositional positions, extended, productless processes, and circulations—a different logic altogether that would suggest the irrelevance of binary categories except as for ideological support.
5. Deploying, then, a performative version of DNA to explain identity, history, gender, and ethnicity becomes a way to perpetuate the cultural binaries of gender and essentialized versions of ethnicity as the stable truth and founding condition of biology. This occurs within and in the face of immensely complex systemic accounts of biochemistry, especially as biochemistry has always tended toward the system and despite the contemporaneous development and availability of systems theories.

This is all, however, not only about the fear of and resistance to change. It is motivated positively by the way it also masks a convenient shift in the nature of the commodity—in our understanding of what there is to make, buy, and sell. DNA, as the combination of information and physical object, becomes both the perfect commodity and the model of a commodity system increasingly dependent on information commodities and subscription systems.

DNA for Sale

A seller on eBay offered his DNA for sale for a starting bid of $11,000.

> I'm selling my DNA, exclusively, giving up my rights forever.
>
> My DNA may lead to a money-making scientific breakthough and I'm offering it to the drug industry.
>
> Item features: The benefits of my DNA are yet to be discovered. The specifications of my DNA are yet to be sequenced. The parts include a very large molecule, made up of smaller units called nucleotides that are strung together in a row. Each nucleotide has three parts: a sugar molecule, a phosphate molecule, and a structure called a nitrogenous base. The nitrogenous base is the part of the nucleotide that carries genetic information. The bases come in four varieties: adenine, cytosine, guanine, and thymine.
>
> Item condition: The item is used. Flaws or repairs are yet to be determined.
>
> Payment and shipping: Buyer must send a trained representative with a buccal swab DNA extraction kit to collect my DNA.[6]

In the late 1980s the U.S. government began the Human Genome Project. Instigated by the Department of Energy but enthusiastically shepherded by James Watson, the Human Genome Project initially became the responsibility of the National Institutes of Health. As Robert Cook-Deegan painstakingly describes it, the project involved complex politics of interagency competition as well as efforts made by the National Research Council and the National Academy of Sciences to define and suggest organizations for the endeavor.[7] As envisioned at its start in 1990, the Human Genome Project was a nonprofit public project aimed at garnering both knowledge and sequencing technologies. The project was divided into two phases: a mapping phase, scheduled for the first five years, in which the locations of various known genes on the twenty-three chromosomes were determined as landmarks; and a sequencing phase when the actual sequences would be worked out. But very quickly, the grant process divided the genome up into twenty-three parts, each of which was claimed by different grant-seeking laboratories. Aware of possible inefficiency and errors, the initial six participating centers, all nonprofit institutions, committed to producing highly accurate results. In addition, they wanted the project's results to be freely available and in the public domain, a commitment they affirmed at a conference in 1996.

In 1998 in the face of lagging results from nonprofit labs and the immense value of the information being gathered, Craig Venter joined with PerkinElmer to begin a privately funded, speedy sequencing project. Using sequencing machines and a whole genome shotgun approach in which the careful mapping of the HGP's first five years became irrelevant, Venter proposed using the sequencing machine first to sequence and then afterward using computer programs to map. As a scientist at the NIH in the early 1990s, Venter had devised a method for sequencing "expressed sequence tags" (EST), short sequences of DNA that identified the presence of a gene. Advised to patent these ESTs because of their potential value, Venter proceeded with patent applications in the name of the NIH. After leaving the NIH, Venter formed The Institute for Genomic Research (TIGR), a nonprofit institute with a symbiotic relation to the for-profit company Human Genome Sciences. Throughout the 1990s, Venter worked on devising faster ways to identify genetic information, giving HGS first dibs for commercial development before releasing the information to the scientific community.[8]

Venter's venture with PerkinElmer in 1998, then, was only the continuation of a practice of scientific commercialism based on the theory that the profit motive provides a greater impetus for quickly sequencing the human genome than the awkward organization of the NIH and its competing satellite labs. But what did PerkinElmer think it was going to get out of the human genome? What profit is there in a genome?

Projects like Venter's produced large databases of genetic information. Selling the right to access such information was a way to produce profit. If, as HGS had already, a gene-sequencing company sold a pharmaceutical company the right to see its information before that information was released to the general public, the pharmaceutical company could discover information of value to its ongoing development of drugs without having to pay licensing rights to patent holders. Subscriptions to the informational database were valuable commodities. PerkinElmer devised the same scheme with the added profit-making side venture of actually manufacturing the gene-sequencing machines. Venter's new company, named Celera, was conceived by Venter as the best of both worlds: a free database of the human genome with a sliding-scale information base offering detailed information on genetic variations and animal genomes. Venter's boss, the PerkinElmer CEO Tony White, was willing to invest $300 million on the bet that pharmaceutical companies would be willing to pay big bucks for the information Celera would generate.

The model of genome information as a commodity not only depends on the momentous decision to make DNA information patentable in certain forms, it also depends on a shift in the way the commodity itself works. If we understand a commodity to be an object of exchange representing a certain amount of labor and capital investment, then the commodity represents an economy of productive joinder, and the more or less contemporaneous trade of value up and down a chain of discrete sites (from vendors to suppliers to manufacturers to middlemen to retail to consumer). This is still close to the commodity culture described and critiqued by Karl Marx and Marxists, one that has evolved but not discernibly altered the notion of the commodity that in itself represents the meeting of labor and capital, desire and use value, fetish and excess.[9] Post-capitalist critiques focus on the extended distribution of production sites, the increasingly fragmented process of production, and the global movements that perpetuate the inequalities between labor forces and capital as well as increasingly alienating the relation between work and product.

When information becomes a way to embody and store labor as it does in a postdigital age of massive searchable records, information also becomes a commodity. But information commodities hide labor within the compensatory strategies of interfaces (such as those enabling us to use computers). Communications technologies, for example, fool us by appearing, at least initially, as products—as telephones, radios, televisions, computers, PDAs. Quickly, however, it became apparent that the real commodity in communications was not the machine (though it is still a commodity indeed) but access to networks. While the machines still require purchase and must be updated periodically, promising a continuous market for technology, access to networks is never ending. As a commodity, such access is anaclitic and perpetual—a subscription in the model of addiction. Configured as a utility, access is an indenture, a charge for each month of "connected" living, producing an ever-increasing dependence.

The anaclitic nature of access commodities is linked to how discrete instances of information appear to be the actual commodity purchased. With cable television, for example, we don't think we are renting access, we think we are purchasing programming. With Internet access, we don't think we are buying access, we think we are purchasing the Internet's resources. In fact, we are purchasing the right to peruse—the right to shop. Commentators such as Katherine Hayles have shown how information itself has become a commodity, and it has indeed become capital. The consumer even rents access to commercials.[10]

The model of information as a capital investment spurred PerkinElmer to rent access to its databases to corporations. In this kind of scheme, its patents on genomic information become capital. The race to complete the human genome was partly spurred by the value of genomic information as a commodity for the private entities engaged in the project. Celera patented as much information as it could in the hopes that its proprietary rights in this information would prove to be a valuable future commodity. Such capital, however, becomes a commodity only in situations where capital itself is never expended. In other words, the kind of commodity information capital can become is a commodity of access and potential opportunity, not something that is itself sold whole cloth. If the information itself is sold or is in the public sector, there is nothing left to rent.

In this way, data banks become commodities by subscription. Instead of purchasing a law library, for example, a law firm purchases a subscription to LexisNexis. Instead of buying *Encyclopaedia Britannica,* a home consumer purchases a subscription right to access. Instead of buying antiviral software that comes with free updates, consumers purchase yearly subscriptions to protection services. Data banks of DNA information—the human genome, the genomes of various research organisms such as mice and roundworms—are rented, updated subscriptions. Such "banks" are not accidentally named as such; they are repositories of capital, but they exist in no specific locatable site and are not static, countable, or like invested money, self-multiplying.

The imaginary of the genetic information commodity arguably spurred the significant public investment in its production. The imaginary character of this commodity was a "book," a tangible, moveable, usable resource, like something in the reference room of a library. And the notion of the genome as an information commodity is still currently confused with a more traditional notion of commodity object, including DNA models, DNA apparatus (one can now buy DNA sequencers on eBay), DNA home test kits, and DNA testing services, or individual genomes (unmapped, of course). There are more than six available online DNA testing vendors who offer DNA tests from paternity and sibling tests to banking and child security kits for prices ranging from $130 to $1,035, depending on the number of subjects and tests involved. One can also buy bird-sexing tests and, of course, one's links to the world family tree. Pharmaceutical companies such as Genaissance Pharmaceuticals offer agricultural genotyping services, a wide range of diagnostic and therapeutic products, databases, and drug development support.[11] While these commodities are offered to other research and pharmaceutical entities, the company itself seeks investors.

The Internet is the primary site for offering such services, at least for the typical nonbusiness consumer. But like the model of international cooperation by which the human genome was mapped, products, services, and databases associated with information banks and sold on the Internet are transnational, floating, often unlocated. Although one sends one's specimens to specific addresses, other services like billing and communications may emanate from anywhere. Like the ideal of the genome—which is an ideal of an unlocated everyman distilled into an information bank erected for permanent subscription—DNA information commodities model transnational corporatism even as the commodity evaporates into a scheme of perpetual siteless rental.

The commodity of genetic information exists partly on the basis of another dream—that someday we might know what to do with it. Its status is that of a speculative toll road waiting for travelers who might pay to use it. And this is perhaps its more dangerous suggestion. As the information commodity shifts the market system from investment to subscription, the commodity itself shifts from present exchange to a perennial indenture, asking us to "pay it forward" indefinitely. Like credit debt, information technologies and especially the information commodity offered by the human genome, whose uses and value we expect to know sometime, appear to offer something for payment now and in the future, a commodity whose future value we can only guess at. This finally makes evident the other faces of capital investment: gambling and paying forever for the same unexpended commodity, hoping that somehow, someday, someone will win the jackpot.

DNA, a small operative molecule, becomes the twirling model of an imaginary of self-replicating wealth, a cosmos offering the ideal of perpetual payment for self-replicating information. It is the commodity culture dream of capital: a product that makes itself and produces continuous income. We see this imaginary in other scenarios: the idea of a machine culture in which machines make machines. Those ideas, however, unlike DNA, at least in science fiction films such as *The Terminator* or *The Matrix* (themselves apparently self-replicating), always result in the threat of human destruction. DNA is not the cause, but its figurations enable a seamless, motivated transition from the possibility of the illusions of ownership and material freedom to a scheme of perpetual indebtedness. All lines—the distinctions among organisms, concepts of material property, the meaning of political and social differences—for good or bad, disappear in the service of an all-subsuming bondage holding the promise of immortality.

The Art of the Hormone

If DNA enables a shift to different ways of thinking about commodities while preserving and making operatively evident more conservative forms of thought such as stable, kinship-based identity, what ways of thinking is it eliding and what might these hidden logics change? Commodity systems have moved from their more conservative structural formula of capital and labor into a networked, dislocated, multileveraged cybernetic system that can systematically offer information as a product. The commodity object is displaced (or replaced) by the subscription, by the purchase of expiring rights.

In biology, what this pseudoscientific performative version of DNA elides is any system that operates between DNA and tissues—the operations of RNA, the production of proteins, the interrelations of genes, the operations of hormones. These processes are all systemic and cybernetic in that they are complexly interdependent, work around a series of balancing system regulators, and demonstrate how there is not necessarily a discrete and stable relation between DNA chains and a constant signification. Some of this process becomes visible in sites of transformation where both structural and systems approaches become simultaneously visible. One site is the occulted "other" scene of gender production, which we might call the "art of the hormone," which tends to be sophomorically literal. Not only do these hormonal sites bring the efficacies of instant and performative action into question—particularly accounts of gender as performative—but they also demonstrate the extent to which biological processes are never really performative or instant at all: their intervening systematicity is one of the elided parts of the DNA/gene story. These sites also suggest finally how the account of gender itself as performative, which appeared in the early 1990s, is the premiere and completely logical symptom of the elision of complexity in the name of will, but which also, alas, is a means of conserving familiar dialecticism.[12]

Let us return to our comic science films from 1952, 1963, and 1996, all three filmic versions of nearly the same scenario. A geeky, insufficiently masculine protagonist concocts a formula, which when imbibed produces a sometimes anguished series of alterations resulting in a juvenated, highly masculine hero. While drawing, at least in the case of Jerry Lewis, from the fount of images surrounding Dr. Jekyll and Mr. Hyde, the transformation these three films render are all processes—a series of stages and recognitions, angst, destruction, and unconsciousness, and finally what we call these days "the reveal"—the newly conformative exemplar of masculinity meets maleness, as our protagonist occasions stares and admiration in new clothes. In relative terms, these transformations seem almost immediate, though in the case of the *Nutty Professor* (both versions), they occur after a series of makeover attempts in the gym. Both situate biochemistry as an efficacious intervention into a story gone wrong, that is, the protagonist isn't boy enough and hasn't enough confidence to get the girl. But what all three also present is an account of the hormonal, a speeded-up version of a hormonizing that transforms geeks into manly irresistible men.

Cary Grant's transformation in *Monkey Business* goes by in discernible, rather leisurely stages: he feels dizzy, realizes he no longer needs his glasses, tests his joints, buys a new wardrobe and sports car, and goes out with Marilyn Monroe. His style is less epiphanic than measured, probably because he has less distance to go to be the epitome of manhood, but also because his alteration is less willed than accident. Lewis's transformation in *Nutty Professor* is an extended clutching and groveling through the violent spill of primary colors and broken glass (transformations require maximal gratuitous breakage), all accompanied by the resonantly Poe-like prophecies of his pet raven, cautioning him to think. His teeth, bad to begin with, get even worse, and he turns into something like a werewolf before he, too, at this point out of our vision, finally transforms to manliness, buys a new wardrobe, and emerges to the dumbfounded stares of passersby and the stunned admiration of the Purple Pit's nightclubbers. Eddie Murphy's transition in *Nutty Professor* is quicker, gauged through the computer to which he is attached. Professor Klump, too, writhes and breaks glass. He too transforms into manliness out of our sight, the change signified by the empty lab coat, his lovely symptomatic exclamations over the disappearance of feminine traits such as his titties and his sighting of his own penis. He too engages in postglandial clothes shopping. All three, however (Grant least spectacularly—though all spectacularly lose their spectacles as they gain potency, an oedipal nugget I don't need to comment further on), undergo a process of gendering that is neither instant nor invisible. It is not, in other words, performative at all. Instead, it makes visible as fantasy the crucial terms of gender as an instantly efficacious performative, subjective will, and gender archetypes, but by inserting another anguished, less-controlled process in between.

The idea of gender as performative arose around 1990, promulgated famously by Judith Butler in *Gender Trouble.* Butler's idea, extensively cited and largely misunderstood, is that subjects acquire a gender by performing it. Acting like a gender makes one that gender. As she describes it, "Gender ought not to be construed as a stable identity or locus of agency from which various acts follow; rather gender is an identity tenuously constituted in time, instituted in an exterior space through a *stylized repetition of acts.* The effect of gender is produced through the stylization of the body and, hence, must be understood as the mundane way in which bodily gestures, movements, and styles of various kinds constitute the illusion of an abiding gendered self."[13] Acting like a man makes the man: gender is not innate but historical and accruing. It is, like DNA, all in the style. But somehow what is really a process of imitation becomes, a few paragraphs later, an effect of performativity. Butler goes on: "If gender attributes, however, are not expressive but performative, then these attributes effectively constitute the identity they are said to express or reveal."[14] The stylized repetition of acts turns into the performative production of gender identity. Performance becomes performative, not "performative" as an adjective referring to a genre involving performance but performative as an analogy to a category of efficacious language or DNA in its pseudoscientific version.

This slippage from performance to the performative is the symptom. Not only does it in itself represent an instance of the same "style" of elision and instantaneity so admired in the discernment of DNA's structure, it also, like accounts of genetic action, short-circuits the historical process—and all of the other variables and processes involved in a complex scenario of engenderment. If, on its face, gender is a repetition of style, even that formulation hides the complex system of events involved in the production of an always contingent subject, the styles available, the consciousness or unconsciousness of the process, and so forth. At the same time, as with accounts of DNA, shifting to the performative means doing the following. First, the complex systems involved that do not constitute a satisfying story at all are masked.

Second, in eliding all stories, the performative implies and depends on the default oedipal gendered narrative—kind of an invisible bait and switch. Third, one implication of the oedipal narrative is the conscious will of the subject/protagonist. Another implication is the binary gendered nature of all choices. The net effect of this is the transformation from what might be an open system of multiple and changing genders to the implied narrative of a willing and conscious subject who chooses his or her gender from the two styles available, then dons that gender over and over. The performative, in other words, is about producing the illusion of power in an environment of its impossibility. Even if Butler is not actually saying this, the passages I have cited have been taken this way by a generation of critics who see performative gender as a form of liberation.

But gender, as Butler suggests, is a species of work, and the work of gender is neither the work of repetition nor the product-oriented work defined by our narratives of production. It is, instead, the work laid out by the economies and mechanisms of the systems first elided by the evocation of performativity. As we all know, DNA is already a synecdoche for genes that are themselves a synecdoche for an incredibly complex system of production, regulation, and multiple causality that works more like a system than performatively, more like a complex machine than a text or even a computer. Gender, too, is already multiple, undertaken, shifting, contextual, adjusted, never stable or the same. This instability is moored by the comforting twoness of the choices on forms, the clothing departments in stores, laws, mores, kinship systems.

We all know that by themselves genes do not produce sex or gender. Genetic females might appear phenotypically as males and vice versa, often as a result of the actions of hormones during fetal development.[15] Even if genes produce hormones as well, the production of phenotypic sex and gender traits is due to far more complex interactions. In other words, XX does not directly and unequivocally equal F (and not even always that) without a lot of work. Not to mention that there are many versions of F or M or something in between or outside or somewhere else we don't even acknowledge. Many systems intervene between genes and genders, but at this point hormones—specifically steroids, the class of hormones to which sex hormones belong, might demonstrate the logic that a pseudoscientific performative notion of DNA elides.

Steroid hormones work through a feedback system. Because they are hydrophobic, that is, they do not dissolve easily in water, they diffuse into cells instead of remaining on the cell wall as some other hormones do. Inside cells that contain protein receptors for the hormone, the hormone binds to the receptor—to areas of DNA responsive to the hormone. This whole complex—DNA receptor and hormone—"transcribes" a part of the DNA that turns other areas of the gene on or off. How much is turned on or off depends on how much hormone actually circulates to incite transcription. How much hormone circulates is governed by three homeostatic mechanisms. One mechanism is a simple feedback machine.[16] One hormone stimulates the production of another whose production then suppresses the continued production of the first. The second mechanism is through the coincidence of antagonistic hormones—hormones that do the opposite of one another—that is, one causing the level of glucose in the blood to rise (glucagon), the other causing it to fall (insulin). A third mechanism is the reaction to the rising or falling levels of another substance. This complex regulatory system is not at all performative, nor can it be analogized to language, oedipal narrative, computers, or other representational or immediately efficacious styles. Instead, it is the essence of process, gradual and balanced change, perpetuated work in which causes and effects abound as part of an extenuated process rather than as beginning and end to discrete stories. Its complexity invites pseudoscientific glosses that simply link substance to behaviors: estrogen to emotionality, testosterone to aggression.

The film transformations, then, even in their abbreviated form, demonstrate in their rendering this anguished interval of work in their introjection of a chemical substance that produces a series of struggles, compensations, and gradual alterations, curiously like a speeded-up version of puberty, like a rapid hormonal transformation. Although two of the three transformations are situated as an effect of the subject's will, all present the coming to better gender as both physical and social. The nutty professor protagonists have calculated the hormonelike effects of their cocktails. The formulas intervene to produce a changed story, better masculinity, alterations in wardrobe and social skills, the emergence of potency and sexual confidence, and, alas, ephemerality as the potions wear off and the protagonists return to their previous, far less stereotypically masculine conditions.

Clearly about improving gender rather than acquiring it, the hormonelike potions work in a *Wizard of Oz* narrative, when all three protagonists realize that their previous states of being were just fine. Even though the Eddie Murphy professor dallies with genetic reconstruction, the genetic basis for these characters' gender is not in question. What their chemically induced transformations illustrate instead is the elided link between genotype and phenotype, the entire system of mechanisms including hormones that build, maintain, and regulate body processes. And though that revelation is still an elision that renders complex processes neatly automatic and superficial, what the scenes point toward is the missing complexity that exists between gene and body, desire and fulfillment.

If, in fact, we actually learned to read—actually understood that language is multivalent, that nothing exists in a stable, secure relation—our abilities to understand and deploy substances such as DNA would in the end be much greater. Scientists involved in genetic research in all of its manifestations know this, but a largely undereducated public continues to digest its easy meals of pseudoscience nuggets that mate past and future in a commodified present. The final effect is a perpetual temporal disjunction, a feeling that the future is now and yet we haven't gotten there, that much publicized advances are disappointing—don't live up to their hype, yet they have not yet been realized. The future disappoints before it even gets here. But with the comforts of DNA we no longer need to puzzle over the mysteries of being or even identity. Even if we don't know the answer we know the answer. DNA has become doctrine, belief, key in all of its unspecified, homuncular, fairy tale glory.

Notes

1. The Epic Acid

1. James Watson and Francis Crick published their first article outlining the structure of DNA, "The Molecular Structure of Nucleic Acids: A Structure for Deoxyribose Nucleic Acid," in *Nature* 171, no. 4356 (April 25, 1953): 737–38. J. L. Austin's lectures that formed the basis of his book *How to Do Things with Words* (Cambridge, Mass.: Harvard University Press, 1975) were delivered in 1953.

2. The history of the discovery of DNA's structure is related by both Watson in *The Double Helix* (New York: Penguin, 1969); and Crick in *What Mad Pursuit: A Personal View of Scientific Discovery* (New York: Basic Books, 1990), as well as by such other historians as Lily E. Kay, *Who Wrote the Book of Life? A History of the Genetic Code* (Stanford, Calif: Stanford University Press, 2000); and Horace Freeland Judson, *The Eighth Day of Creation: Makers of the Revolution in Biology* (Plainview, N.Y.: CSHL Press, 1996).

3. Richard Dawkins, *The Selfish Gene* (London: Oxford University Press, 1990), 32.

4. Richard Lewontin, *It Ain't Necessarily So: The Dream of the Human Genome and Other Illusions* (New York: New York Review of Books, 2000), 193.

5. Ibid., 104.

6. Ibid., 193.

7. Ruth Hubbard and Elijah Wald, *Exploding the Gene Myth: How Genetic Information Is Produced and Manipulated by Scientists, Physicians, Employers, Insurance Companies, Educators, and Law Enforcers* (Boston: Beacon, 1997).

8. Ibid., xiii.

9. Lewontin, 197.

10. Michel Morange, *The Misunderstood Gene,* trans. Matthew Cobb (Cambridge, Mass.: Harvard University Press, 2001).

11. Ibid., 22.

12. Ibid., 22–23. See www.oed.com.

13. Ibid., 23.

14. Ibid., 24.

15. Jean-François Lyotard, *The Postmodern Condition: A Report on Knowledge,* trans. Geoff Bennington and Brian Massumi (Minneapolis: University of Minnesota Press, 1984).

16. Evelyn Fox Keller, *Refiguring Life: Metaphors of Twentieth-Century Biology* (New York: Columbia University Press, 1995), 3.

17. Evelyn Fox Keller, *The Century of the Gene* (Cambridge, Mass.: Harvard University Press, 2000).

18. Ibid., 40.

19. Erwin Schrödinger, *What Is Life?* (Cambridge: Cambridge University Press, 1992).

20. Keller, *Century,* 5.

21. Ibid., 72.

22. Ibid., 130.

23. Ibid.

24. Ibid., 136.

25. Richard Doyle, *On Beyond Living: Rhetorical Transformations of the Life Sciences* (Stanford, Calif.: Stanford University Press, 1997), 26.

26. Ibid., 28.

27. Ibid.

28. Dorothy Nelkin and M. Susan Lindee, *The DNA Mystique: The Gene as a Cultural Icon* (New York: Freeman, 1995), ix.

29. José van Dijck, *Imagenation: Popular Images of Genetics* (New York: New York University Press, 1998), 3.

30. Ibid., 4.

31. Celeste Condit, *The Meanings of the Gene: Public Debates about Human Heredity* (Madison: University of Wisconsin Press, 1999), 10.

32. Understandings of narrative as a version (or pattern) for hetero-reproduction include Robert Scholes and Robert Kellogg, *The Nature of Narrative* (New York: Oxford University Press, 1968); Teresa de Lauretis, *Alice Doesn't: Feminism, Semiotics, Cinema* (Bloomington: Indiana University Press, 1984); and Judith Roof, *Come as You Are: Sexuality and Narrative* (New York: Columbia University Press, 1996).

2. Genesis

1. Nonstructuralist modes of thought are many and complex. They tend to share several qualities: some aspect of relative perspective; some form of cybernetic or feedback mechanism or interdependency; and some species of multiple simultaneity. If Einstein envisions space and time as functions of one another, and Richard Feynman contrives an understanding of particle movement as a sum over the histories, or the sum of the statistical possibility of all of the paths a particle might take, or Norbert Wiener thinks about the ways systems might regulate themselves, none of these is either linear or subject to any simple cause-effect mechanism or even hierarchy (though systems may have hierarchies). Thinking about phenomena as complex, intertwined modes irreducible to any single basic element or cause evades reductionism.

2. This broad presentation of conceptual history comes from John L. Heilbron, *The Oxford Guide to the History of Physics and Astronomy* (New York: Oxford University Press, 2005); Geoffrey E. R. Lloyd, *Early Greek Science: Thales to Aristotle* (New York: Norton, 1974); and David C. Lindberg, *The Beginnings of Western Science: The European Tradition in Philosophical, Religious, and Institutional Context* (Chicago: University of Chicago Press, 1992).

3. Sir Isaac Newton, *Optics, Modern History Sourcebook*, http://www.fordham.edu/halsall/mod/newton-Optics.html (accessed February 28, 2006).

4. Aristotle, *Categories*, in *Aristotle: Selections*, ed. W. D. Ross (New York: Charles Scribner's Sons, 1955), 5. See also James G. Lennox, *Aristotle's Philosophy of Biology* (New York: Cambridge University Press, 2001).

5. Aristotle, *Metaphysics*, in *Aristotle: Selections*, 54.

6. This is obviously a scanty treatment of a large chunk of the history of science. The Renaissance not only accomplished the revival and translation of the works of Greek philosophers and physicians but reinterpreted the material according to its own fascination with the body. For a summary of

Renaissance medicine, see Roy Porter, *The Greatest Benefit to Mankind: A Medical History of Humanity* (New York: Norton, 1997), 163–200.

7. Quoted in Porter, 224.

8. Quoted in Porter, 223.

9. Francis Bacon, *Novum Organum, Internet Encyclopedia of Philosophy,* http://www.utm.edu/research/iep/b/bacon.htm, 363–64 (accessed February 28, 2006).

10. Albert Einstein, "Über einen die Erzeugung und Verwandlung des Lichtes beteffenden heuristischen Gesichtspunkt," *Annalen der Physik* 17 (1905): 132–48.

11. The modern history of atomic theory is traced by Bernard Pullman in *The Atom in the History of Human Thought,* trans. Axel R. Reisinger (New York: Oxford University Press, 2001). Pullman's history also shows the extent to which many cultures attempted to suppress the idea of the atom.

12. Obviously, the physics of subatomic particles is infinitely more complex than this, but further information on M theory, string theory, and quarks can be found in Stephen Hawking, *The Universe in a Nutshell* (New York: Bantam, 2001); and Brian Greene, *The Elegant Universe: Superstrings, Hidden Dimensions, and the Quest for the Ultimate Theory* (New York: Vintage, 2000).

13. Greene, 17.

14. The history of biochemistry is exhaustively outlined by Horace Freeland Judson in *The Eighth Day of Creation: Makers of the Revolution in Biology*. For a more philosophical account of this trajectory, see part 3 of Georges Canguilhem, *A Vital Rationalist: Selected Writings from Georges Canguilhem,* ed. François Delaporte, trans. Arthur Goldhammer (New York: Zone, 2000).

15. Ferdinand de Saussure, *A Course in General Linguistics,* trans. Roy Harris (New York: Open Court, 1986).

16. Alexandre Kojève, *Introduction to the Reading of Hegel: Lectures on the Phenomenology of Spirit,* trans. Raymond Queneau (Ithaca, N.Y.: Cornell University Press, 1980). Kojève glosses much of Hegel's thought. The material quoted here is from Kojève's first chapter, reproduced at http://www.marxists.org/reference/subject/philosophy/works/fr/kojeve.htm.

17. Kojève, http://www.marxists.org/reference/subject/philosophy/works/fr/kojeve.htm.

18. Sigmund Freud, *The Interpretation of Dreams,* ed. and trans. James Strachey (New York: Avon, 1965), 343.

19. Jacques Lacan, "The Insistence of the Letter in the Unconscious," quoted in *Jacques Lacan and the Philosophy of Psychoanalysis,* by Ellie Ragland-Sullivan

(Urbana: University of Illinois Press, 1986), 236. I cite this instead of the version translated by Alan Sheridan, as Ragland-Sullivan's translation is clearer in this instance.

20. Jacques Lacan, *Four Fundamental Concepts of Psychoanalysis,* trans. Alan Sheridan (New York: Norton, 1981), 149.

21. Ibid.

22. Ibid., 150.

23. Ibid.

24. Ibid.

25. Ibid., 20.

26. Ibid., 151.

27. This work is collected in Claude Lévi-Strauss, *Structural Anthropology,* trans. Claire Jacobson and Brooke Grundfest Schoepf (New York: Basic Books, 1963).

28. Ibid., 71.

29. Ibid., 72.

30. Ibid.

31. Ibid., 73.

32. Roland Barthes, *Mythologies,* trans. Annette Lavers (New York: Noonday, 1972).

33. Roland Barthes, *Elements of Semiology,* trans. Annette Lavers and Colin Smith (New York: Hill and Wang, 1967), 23.

34. Ibid., 24.

35. Ibid.

36. Ibid., 25.

37. Roland Barthes, "The Structuralist Activity," in *Critical Essays,* trans. Richard Howard (Evanston, Ill.: Northwestern University Press, 1972), 214.

38. Ibid., 215.

39. Ibid.

40. Barthes, *Mythologies,* 129.

41. Barthes, "Structuralist Activity," 142.

42. Ibid.

43. Ibid., 142–43.

44. Austin, 6.

45. Jacques Derrida, *Limited, INC* (Evanston, Ill.: Northwestern University Press, 1988).

46. Schrödinger, 3.

47. Ibid., 61.

48. James D. Watson, *A Passion for DNA: Genes, Genomes, and Society* (Cold Spring Harbor, N.Y.: Cold Spring Harbor Laboratory Press, 2000), 5.

49. Schrödinger, 48.

50. Ibid.

51. Ibid., 22.

52. See also Doyle's account of Schrödinger's rhetoric in chapter 2 of *On Beyond Living*.

53. Watson, *Passion for DNA*, 123.

54. Michel Foucault, *The Archaeology of Knowledge*, trans. A. M. Sheridan Smith (New York: Pantheon, 1972), 208.

55. Ibid., 100.

56. Ibid., 114.

57. Ibid.

58. Ibid., 183.

59. Ibid., 185.

60. Norbert Wiener advanced his ideas on cybernetics in the late 1940s. See his later edition *Cybernetics, or The Control and Communication in the Animal and the Machine*, 2nd ed. (Boston: MIT Press, 1965).

61. Heisenberg discerned that "the more precisely the position is determined, the less precisely the momentum is known." Further, the implications of this uncertainty on the study of physics produced an acknowledgment of the effect of the observer as a part of the phenomenon being observed. "I believe," Heisenberg states, "that the existence of the classical 'path' can be pregnantly formulated as follows: The 'path' comes into existence only when we observe it" ("Uncertainty Principle Paper," 1927, 478–504, *AIP Center for History of Physics*, http://www.aip.org/history/heisenberg/bibliography/1920-29.htm [accessed February 28, 2006]).

62. Gregory Bateson, *Steps to an Ecology of Mind* (Chicago: University of Chicago Press, 1972), 490.

63. Ibid.

64. Weiner, 6–7.

65. Ludwig von Bertalanffy, *General System Theory* (New York: Braziller, 1968), 5.

66. Ibid., 13.

67. Ibid., 6.

68. Anthony Wilden, *System and Structure: Essays in Communication and Exchange*,

2nd ed. (London: Tavistock, 1972); Niklas Luhmann, *Ecological Communication,* trans. John Bednarz Jr. (Chicago: University of Chicago Press, 1989); Gilles Deleuze and Félix Guattari, *A Thousand Plateaus: Capitalism and Schizophrenia,* trans. Brian Massumi (Minneapolis: University of Minnesota Press, 1987).

69. Cary Wolfe, *Critical Environments: Postmodern Theory and the Pragmatics of the "Outside"* (Minneapolis: University of Minnesota Press, 1998), 53.

70. Thomas Kuhn, *The Structure of Scientific Revolutions,* 3rd ed. (Chicago: University of Chicago Press, 1996), 92.

71. Von Bertalanffy, 13.

72. Watson, *Passion for DNA,* 162–63.

3. Flesh Made Word

1. Watson, *Double Helix,* 125.

2. Ibid., 126.

3. Watson and Crick, "Molecular Structure of Nucleic Acids," 737, quoted in Judson, 170. Judson's book contains facsimiles of both of Watson and Crick's 1953 articles on DNA.

4. Ibid.

5. J. D. Watson and F. H. C. Crick, "Genetical Implications of the Structure of Deoxyribonucleic Acid," *Nature* 171, no. 4361 (May 30, 1963): 964–67, quoted in Judson, 171.

6. Quoted in Judson, 155.

7. The protagonist of James Joyce's *Stephen Hero* (New York: New Directions, 1959), describes an epiphany: "This is the moment which I call epiphany. First we recognise that the object is one integral thing, then we recognise that it is an organised composite structure, a thing in fact: finally, when the relation of the parts is exquisite, when the parts are adjusted to the special point, we recognise that it is the thing which it is. Its soul, its whatness, leaps to us from the vestment of its appearance. The soul is the commonest object, the structure of which is so adjusted, seems to us radiant. The object achieves its epiphany" (99). The epiphany is an uncanny description of the status of DNA at the identification of its structure.

8. In an interview on the *Nova* segment "Cracking the Code of Life," first aired on PBS on April 17, 2001, Lander characterizes the genome as a "parts list" in the face of the more hyperbolic claims about the "book of life." See http://www.pbs.org/wgbh/Nova/genome/ (accessed November 30, 2004).

9. Gregor Mendel, "Experiments in Plant Hybridization," *Journal of Heredity* 42, no. 1 (1951): 3–47. This is a reprint of two lectures Mendel delivered on February 8 and March 8, 1865, at the Naturforschedenden Vereins of Brünn. The paper is available at http://www.mendelweb.org/Mendel.html (accessed February 28, 2006). Mendel's ideas were simultaneously rediscovered by Hugo deVries, Carl Correns, and Erik von Tschermak. William Bateson enthusiastically embraced them in his 1900 paper, "Problems of Heredity as a Subject for Horticultural Investigation."

10. Bateson used the term *genetics* in a letter. The Web site for TIGR: The Institute for Genomic Research provided a time line in 2002 that listed the following comment: "Four years earlier, William Bateson, an early geneticist and a proponent of Mendel's ideas, had used the word genetics in a letter; he felt the need for a new term to describe the study of heredity and inherited variations. But the term didn't start spreading until Wilhelm Johannsen suggested that the Mendelian factors of inheritance be called genes." The time line is no longer available, but see generally http://gnn.tigr.org/.

11. University of Cambridge, Department of Genetics, "A Brief History of the Department," 2002, www.gen.cam.ac.uk/dept/dept_history.html (accessed February 28, 2006).

12. The *American Naturalist* is cited in the entry on "gene" at www.oed.com. Johannsen's terminology lays out the basic parameters of the field of genetics. For a complete discussion of his terms see Frederick Churchill, "Wilhelm Johannsen and the Genotype Concept," *Journal of the History of Biology* 7 (1974): 5–30.

13. See, for example, the history of gene mapping, as set out in Jonathan Weiner, *Time, Love, Memory: A Great Biologist and His Quest for the Origins of Behavior* (New York: Knopf, 1999); and Lily Kay's careful history in *Who Wrote the Book of Life?*

14. Schrödinger, 21.

15. Ibid., 22.

16. Ibid., 61.

17. Ibid., 62.

18. Watson, *Passion for DNA*, 5.

19. Quoted in Judson, 172.

20. Quoted in Judson, 153.

21. Quoted in Judson, 173.

22. See Judson, 256, and Kay, chap. 4.

23. See Judson, 256.

24. Ibid., 248.

25. Ibid., 281.

26. Quoted in Judson, 256.

27. Jeremy Campbell, *Grammatical Man* (New York: Simon and Schuster, 1983); Steve Jones, *The Language of the Genes* (New York: Anchor, 1995); Walter Bodmer and Robin McKie, *The Book of Man* (London: Oxford University Press, 1997); and Nicholas Wade, *Life Script: How the Human Genome Discoveries Will Transform Medicine and Enhance Your Health* (New York: Simon and Schuster, 2002).

28. Toy and Watson were both quoted multiply on several online news services, including Reuters and AP.

29. Maggie Fox, "First Look at Human Genome Shows How Little There Is," http://dailynews.yahoo.com/h/nm/20010211/ts/genome_dc_2.html (February 11, 2001).

30. Matthew Ridley, *Genome: The Autobiography of a Species in Twenty-three Chapters* (New York: HarperCollins, 1999).

31. Ibid., 6.

32. Dawkins, 22.

33. In *The Interpretation of Dreams,* Freud discerned condensation and displacement as the two logics through which dreams were organized. In his readings of Freud, Lacan adopted Jakobson's understandings of metaphor and metonymy to gloss Freud's insight in linguistic terms. Jakobson saw metaphor and metonymy as two "poles" of organization that were interrelated. See Roman Jakobson, *On Language,* ed. Linda Waugh and Monique Monville-Burston (Cambridge, Mass.: Harvard University Press, 1995).

34. For an account of this shift see Judith Roof, *Reproduction of Reproduction: Imaging Symbolic Change* (New York: Routledge, 1995).

35. Seymour Benzer, a drosophila researcher, began studying how certain behaviors such as reactions to light or internal time clocks were regulated genetically. For a full account see Weiner, *Time, Love, Memory.*

36. See, for example, the *Crime Times* article, "Antisocial Behavior, Executive Function Deficits May Be Linked," *Crime Times* 6, no. 2 (2000): 5, http://www.crime-times.org/97c/w97cp5.htm (accessed February 28, 2006).

37. Eric Schlosser, *Fast Food Nation: The Dark Side of the All-American Meal* (New York: Perennial, 2002).

38. Dawkins, 35.

39. Ridley, 191.

40. The possibility that the mutation known as delta 32 was caused by the black death is explored in "Secrets of the Dead: Mystery of the Black Death," PBS, 2002.

41. I make this argument in an earlier essay, "In Locus Parentis," in *Printing and Parenting in Early Modern England,* ed. Douglas Brooks (London: Ashcraft, 2005).

42. J. David Brook, West Bridgford GBX, David E. Housman, M. A. Newton, Duncan J. Shaw, Banchory GBX, Helen G. Harley, Rhiwbina GBX, Keith J. Johnson, and Glasgow GBX, inventors, "Sequence Encoding the Myotonic Dystrophy Gene and Uses Thereof," Patent Number 5955265, September 21, 1999. For the full patent see The DNA Patent Database, http://dnapatents.georgetown.edu/.

43. "Myotonic Dystrophy, 1992," www.DNAPatent.com (accessed February 28, 2006).

44. U.S. patent law was amended in 1999 to include specifically biotechnological products and to redefine what novelty might mean in a biotechnological context. The section on patentability now reads: "Section 103. Conditions for patentability; non-obvious subject matter. (a) A patent may not be obtained though the invention is not identically disclosed or described as set forth in section 102 of this title [35 *USC* § 102], if the differences between the subject matter sought to be patented and the prior art are such that the subject matter as a whole would have been obvious at the time the invention was made to a person having ordinary skill in the art to which said subject matter pertains. Patentability shall not be negatived by the manner in which the invention was made. (b)(1) Notwithstanding subsection (a), and upon timely election by the applicant for patent to proceed under this subsection, *a biotechnological process using or resulting in a composition of matter that is novel* under section 102 [35 *USC* § 102] and nonobvious under subsection (a) of this section shall be considered nonobvious if—(A) claims to the process and the composition of matter are contained in either the same application for patent or in separate applications having the same effective filing date; and (B) the composition of matter, and the process at the time it was invented, were owned by the same person or subject to an obligation of assignment to the same person" (35 *USC* §103 [emphasis added]).

45. Dorothy Wertz defines what is in fact patented in "Gene Letter: Patenting DNA–A Primer" (*ELSI Research,* www.ornl.gov [accessed February 28,

2006]): "Over 5,000 applications have been filed for United States patents on fragments of human DNA. So far, the Patent Office has granted over 1,500 patents. How is it possible to patent something that only God or nature can make? The answer is that it isn't. Nothing can be patented in its natural state, including DNA in a human body. What is being patented is the process of discovering and isolating in the laboratory certain strings of DNA that were not obvious before. Only the DNA that exists in the laboratory can be patented, not the DNA in humans or animals. The patent covers only the process, not the in situ raw material. Someone else could still 'discover' this DNA using another method not covered by the patent, as long as they used their own database. (If they used the patent holder's database, their 'new' method would fall under the original patent. Few researchers want to create a new database.) U.S. law allows the patenting of whatever is discovered through a new method, along with the patenting of the method itself. (Not all countries permit this.) Hence the seeming paradox of patenting human DNA."

46. And hence HUGO's (Human Genome Organization) concern over the proliferation of what are, practically speaking, gene patents. In "Life in Genetics–What's Happening," Dorothy Wertz reports that "HUGO is asking the PTO to reject the estimated 20,000 patent applications currently pending that do not meet criteria for utility. They are also requesting the PTO rescind any non-qualifying patents already approved—including those based on computer predictions of utility. HUGO recently announced its opposition to the patenting of genetic discoveries with as yet unproven utility" (www.ornl.gov).

47. Jeff Lyon and Peter Gorner, *Altered Fates: Gene Therapy and the Retooling of Human Life* (New York: Norton, 1996), 125.

48. Ibid.

49. E. Richard Gold, *Body Parts: Property Rights, and the Ownership of Human Biological Materials* (Washington, D.C.: Georgetown University Press, 1996), 25.

50. Ibid., 28.

51. 17 *U.S.C.* 102.

52. Human Genome Project Information, http://www.ornl.gov/sci/techresources/Human_Genome/home.shtml. The same site also makes available the following statement on the patenting of DNA sequences: "HUGO (The Human Genome Organisation) is worried that the patenting of partial and uncharacterized cDNA sequences will reward those who make routine discoveries but penalize those who determine biological function or application. Such an outcome would impede the development of diagnostics and therapeutics,

which is clearly not in the public interest. HUGO is also dedicated to the early release of genome information, thus accelerating widespread investigation of functional aspects of genes. This statement explains our concerns."

53. HeLa cells were cultured from cells taken in 1951 from a cervical malignancy in a patient named Henrietta Lacks, a thirty-one-year-old African American woman living in Baltimore. Unlike regular cells, cancer cells seem to be able to divide endlessly. HeLa cells are so prolific and aggressive that not only are they still reproducing, they often take over other cell cultures. Taking and culturing such cells without the patient's knowledge or permission has spurred ethical discussions of such practices. The answer in the Lacks case has been to declare the cells independent single-celled organisms. See The Immortal Cells of Henrietta Lacks Web site. In the 1980 case *Diamond, Commissioner of Patents and Trademarks v. Chakrabarty* 447 *U.S.* 303; 100 *S. Ct.* 2204, the Supreme Court determined that a live, human-made microorganism is patentable subject matter under 35 *USC* § 101, giving wide scope to the term "composition of matter" in the patent statutes.

54. In the context of her discussion of the information metaphor in gene discourse in *Refiguring Life,* Evelyn Fox Keller also arrives at the potency of metaphor as a persuasive figuration. She says, "Still, even while researchers in molecular biology and cyberscience displayed little interest in each other's epistemological program, *information*—either as metaphor or as material (or technological) inscription—could not be contained. In the real world, there was no stopping the circulation of meaning, no cutting of what Lacan calls the circuit of language. In the 1960s, metaphor, not material exchange, provided the primary vehicle for this circulation. In other words, it was the metaphorical use of *information*—as it criss-crossed among these two sets of disciplines, their practitioners, and among their subjects—that provided the principal vector for the dissemination of meaning" (103–4).

55. Fox, February 11, 2001.

56. U.S. Department of Energy, "Genomics and Its Impact on Medicine and Society," 2002, www.ornl.gov (February 28, 2006).

4. The Homunculus and Saturating Tales

1. Ridley, *Genome,* 11.

2. Dr. Seuss (Theodore Geisel), *Horton Hears a Who!* (New York: Random House, 1954).

3. For a thorough account of Darwin's theories of evolution and their aftermath, see Stephen Jay Gould, *The Structure of Evolutionary Theory* (Cambridge, Mass.: Harvard University Press, 2002).

4. Ridley, *Genome,* 209.

5. As demonstrated by de Lauretis in *Alice Doesn't.*

6. Both Sandra Harding and Anne Fausto-Sterling, among others, demonstrate the gender biases of science. See Sandra Harding, *The Science Question in Feminism* (Ithaca, N.Y.: Cornell University Press, 1986); and Anne Fausto-Sterling, *Myths of Gender: Biological Theories about Women and Men* (New York: Basic Books, 1992).

7. Dawkins, 45.

8. Ibid., 2.

9. Ibid., 145.

10. Matthew Ridley, *The Red Queen: Sex and the Evolution of Human Behavior* (New York: Penguin Books, 1993).

11. Ibid., 3–4.

12. Ibid., 13.

13. W. R. Rice and A. K. Chippendale, "Intersexual Ontogenetic Conflict," *Journal of Evolutionary Biology* 14 (2001): 685–93.

14. Alberto Civetta and Andrew G. Clark, "Correlated Effects of Sperm Competition and Postmating Female Mortality," *Proceedings of the National Academy of Science* 97, no. 24 (November 21, 2000): 13162–65 (November 14, 2000).

15. Rice and Chippendale, 690.

16. Ibid.

17. Ridley, *Genome,* 115.

18. Paraphrased in Ridley, *Genome,* 215.

19. Ridley, *Red Queen,* 104.

20. Ridley, *Genome,* 217.

21. Anne Fausto-Sterling, *Sexing the Body: Gender Politics and the Construction of Sexuality* (New York: Basic Books, 2000), 3.

22. Ibid., 101.

23. The day after the announcement that the Human Genome Project was 85 percent complete, the *New York Times* ran an essay in which President Clinton was quoted as saying: "Today we are learning the language in which God created life." Nicholas Wade, "Reading the Book of Life: The Overview; Genetic Code of Human Life Is Cracked by Scientists," *New York Times,* June 27, 2000.

24. Freud, for example, lists homosexual desires as a perversion of the "normal" alignment of sexual aim and sexual object in *Three Essays on the Theory of Sexuality,* trans. James Strachey (New York: Basic Books, 1975).

25. Weiner, *Time, Love, Memory,* 196–98.

26. Ibid., 127.

27. There are a number of Web sites devoted to reporting and discussing the question of the genetic basis of homosexuality (mostly male). See, for example, "The Gay Gene," http://members.aol.com/gaygene/ (accessed December 7, 2004).

28. Wald quoted in an interview with Frank Aquino ("Exploding the Gene Myth," 1993, http://eserver.org/gender/exploding-the-gene-myth.html [accessed July 10, 2003]).

29. For an account of the functions of taxonomy and classification systems see Geoffrey Bowker and Susan Leigh Star, *Sorting Things Out: Classification and Its Consequences* (Cambridge, Mass.: MIT Press, 2000).

30. The history of eugenics beliefs, science, and movements is set out in Ruth Clifford Engs, *The Eugenics Movement: An Encyclopedia* (New York: Greenwood, 2005).

31. Luigi Cavalli-Sforza, *Genes, Peoples, and Languages,* trans. Mark Seielstad (Berkeley: University of California Press, 2000).

32. Ibid., 30.

33. Ashley Montagu, *Man's Most Dangerous Myth: The Fallacy of Race* (New York: Harper, 1942).

34. Byron Spice, "Genetics and Race: Researchers Explore Why Rates of Diseases Vary from One Population to Another," *Pittsburgh Post-Gazette,* May 7, 2002.

35. In an editorial response, "DNA and Geneology: A Worrisome Mix," aired on February 15, 2006, in response to the PBS documentary *African-American Lives,* Karla Holloway pointed out the dangers of assuming that genetic science is free of racist conceptions just because race has been declared not to be genetic. She also raised the question of the uniformity of any racial category.

36. See "Boyd E. Graves, J. D. Discovered the United States' Secret 1971 Special 'AIDS' Virus Flowchart in 1999," http://www.stewartsynopsis.com (accessed November 19, 2004).

37. Alison P. Galvani and Montgomery Slatkin, "Evacuating Plague and Smallpox as Historical Selective Pressures for the CCR5-Delta32 HIV-Resistance Allele," *PubMed,* http://www.ncbi.nlm.nih.gov/entrez/query.fcgi?holding=npg&

cmd=Retrieve&db=PubMed&list_uids=14645720&dopt=Abstract (accessed November 9, 2004).

38. Nelkin and Lindee, 2.

39. This brief history of reproduction comes from Porter, *Greatest Benefit to Mankind* and Jon Heilbron's *Oxford Companion to the History of Modern Science.*

40. In *Beyond the Pleasure Principle,* trans. James Strachey (New York: Norton, 1990), Freud presents an extended argument about the biological necessity for sexual reproduction over other, eventually exhaustive forms such as fission. Using the example of paramecia, Freud uses Wiessman's work to demonstrate that those paramecia who reproduce through fission eventually exhaust their lines.

41. Georges Bataille, *L'Erotisme: Death and Sensuality,* trans. Mary Dalwood (London: Boyars, 1987).

42. Francis Galton, *Hereditary Genius: An Inquiry into Its Laws and Consequences* (London: Macmillan, 1869), 1.

43. Ibid.

44. Ibid.

45. See generally Lyon and Gorner for an account of this attempt at gene therapy.

46. Lyon and Gorner, 110–19; see also "Human Genome Project Information," http://www.ornl.gov/sci/techresources/Human_Genome/medicine/genetherapy.shtml (accessed December 10, 2004), as well as "'Miracle' gene therapy trial halted," http://www.newscientist.com/news/news.jsp?id=ns99992878 (accessed February 28, 2006).

47. Stephen Hawking, *The Universe in a Nutshell* (New York: Bantam, 2001).

48. Donna Haraway, *Simians, Cyborgs, and Women: The Reinvention of Nature* (New York: Routledge, 1991).

49. The Web site sponsored by the U.S. government devoted to the genome has a section on ethical, legal, and social issues (ELSI). See www.ornl.gov/sci/techresources/Human_Genome/home.shtml.

5. The Ecstasies of Pseudoscience

1. This history of ideas about blood is condensed from Porter, *Greatest Benefit to Mankind.*

2. Fee tail male limits the ownership of property to male descendants only. Such a limitation follows the land and is understood as a kind of ownership.

3. The introduction of blood typing in 1900 began the process of identifying the father scientifically, though blood types only narrow the field slightly and worked more often to show who couldn't be the father rather than who was.

4. "Intelligent design" is an idea developed by several researchers that seems to require or even prove that a deity exists because phenomena that exist could come about in no other way. See, for example, Michael Behe, *Darwin's Black Box: The Biochemical Challenge to Evolution* (New York: Free Press, 1998).

5. In their chapter "Sacred DNA," Nelkin and Lindee discuss some of these same topics, including religion, personal identification, and the notion of DNA as the secret of life. Their approach is more keyed to the idea that DNA has acquired the powers of a deity.

6. Carl Sagan, *The Demon-Haunted World: Science as a Candle in the Dark* (New York: Ballantine, 1996).

7. *Abracadabra* is a late Latin word that served only as a magical incantation. Its power was originally derived from wearing the word, whose letters were arranged in an inverted triangle. Wearing the triangle of letters protected the wearer against trouble or disease. With DNA, abracadabra's triangle is reduced to three letters.

8. Behe, 10.

9. Ibid.

10. Sagan, 14.

11. See Lyon and Gorner; also Lori B. Andrews, *Future Perfect: Confronting Decisions about Genetics* (New York: Columbia University Press, 2001).

12. There is some small pressure within the United States for more awareness of the global effects of genetically modifying crops and animals. For example, Kristin Dawkins's pamphlet, *Gene Wars: The Politics of Biotechnology* (New York: Seven Stories Press, 1997) surveys the broad interrelation among biotechnologies, governmental policies, and international treaties.

13. For six years, the European Union banned genetically modified food crops, but lifted the ban in the spring of 2004. But even as the ban was lifted, the EU still required that genetically modified foods be labeled as such. Americans are not so fortunate as to even have an informed choice. If there is not a problem with genetic modification, then what's the harm in telling us? See Dawkins, *Gene Wars,* 34–35.

14. Dawkins, *Gene Wars,* 40.

15. See www.monsanto.co.uk (accessed August 5, 2003).

16. This very brief history of fingerprints comes primarily from http://onin.com/fp/fphistory.html.

17. Francis Galton, "Personal Identification and Description," *Nature* 38 (1888): 173–77, 202.

18. Ibid., 202.

19. Several Web sites explain the CODIS system and its basis in STRs. See www.fbi.gov/hq/lab/codis/index1.htm; www.promega.com; the United States Government Genome site, www.ornl.gov; or Interpol's DNA site, http://www.interpol.int/Public/Forensic/DNA/Default.asp.

20. news.bbc.co.uk/2/hi/health/3019324.stm (accessed August 8, 2003).

6. Rewriting History

1. *Genebase Bionet Builder,* http://www.genebase.com/ (accessed February 20, 2006).

2. Relative Genetics, http://www.relativegenetics.com/relativegenetics/index.jsp (accessed February 20, 2006).

3. http://www.familytreedna.com/default.asp.

4. Niall of Nóigiallach, according to a Trinity College, Dublin, study, may be the ancestor of one in twelve Irishmen. This is according to *FamilyTreeDNA,* www.familytreena.com/matchnialltest.html (accessed February 28, 2006). Family TreeDNA are selling tests to determine if one is one of Niall's descendants.

5. Henry Louis Gates Jr. hosted a PBS series *African-American Lives,* first aired in February 2006, which used DNA and other documentation to trace the origins of black celebrities. See www.pbs.org/wnet/aalives.

6. Ebay offering, 2005.

7. This history is drawn primarily from Robert Cook-Deegan, *Gene Wars: Science, Politics, and the Human Genome* (New York: Norton, 1994).

8. Accounts of Venter, Celera, and the HGP race come from James Shreeve, *The Genome War: How Craig Venter Tried to Capture the Code of Life and Save the World* (New York: Knopf, 2004).

9. As described, for example, in Timothy Bewes, *Reification: Or the Anxiety of Late Capitalism* (London: Verso, 2002); and Kevin Kelly, *Out of Control: The New Biology of Machines, Social Systems, and the Economic World* (New York: Perseus, 1995).

10. See Katherine Hayles, *How We Became Posthuman: Virtual Bodies in Cybernetics, Literature, and Informatics* (Chicago: University of Chicago Press, 1999).

11. See www.genaissance.com.

12. This account comes most famously from Judith Butler, *Gender Trouble* (New York: Routledge, 1991), but was fondly adopted and proliferated by critics, performers, and theorists.

13. Ibid., 140.

14. Ibid., 141.

15. Females with a genetic disorder called congenital adrenal hyperplasia have a condition in which the steroid cortisol is not produced, causing the overproduction of other steroids. Female fetuses with this recessive trait produce too many steroids and often develop genitals that appear to be male, with large clitorises or even penises. Such females are often mistaken for boys at birth. Males with androgen insensitivity syndrome develop as females without internal male sexual organs because the body's tissues are completely insensitive to androgens.

16. Hormone systems are complex, but a simple, clear exposition of human hormones exists at http://users.rcn.com/jkimball.ma.ultranet/BiologyPages/H/Hormones.html.

Index

Judith Roof is professor of English at Michigan State University and codirector of the avant-garde performance group SteinSemble.